188
Advances in Anatomy Embryology and Cell Biology

M.L. López-Gracia and M.A. Ros

Left–Right Asymmetry in Vertebrate Development

With 17 Figures and 7 Tables

Springer

Ph.D. Maria L. López-Gracia
Prof. Dr. Marian Ros

Departamento de Anatomía y
Biología Celular
Facultad de Medicina
Universidad de Cantabria
39011 Santander
Spain

e-mail: rosm@unican.es

ISSN 0301-5556
ISBN-10 3-540-36347-5 Springer Berlin Heidelberg New York
ISBN-13 978-3-540-36347-7 Springer Berlin Heidelberg New York

Springer is a part of Springer Science+Business Media
springer.com

Editor: Simon Rallison, Heidelberg
Desk editor: Anne Clauss, Heidelberg
Production editor: Nadja Kroke, Leipzig
Cover design: WMX Design Heidelberg
Typesetting: LE-TEX Jelonek, Schmidt & Vöckler GbR, Leipzig
Printed on acid-free paper SPIN 11795179 27/3150/YL – 5 4 3 2 1 0

List of Contents

Acknowledgements

This work was supported by grants FP12002-02946 from the Spanish Ministry of Education and Science and grant API 05/27 from the Instituto de Formación e Investigación Marqués de Valdecilla. We are very grateful to Dr. J.L. Ojeda for the SEM cilia pictures and Dr. J.M. Icardo for SEM pictures and critical reading of some parts of this manuscript. The authors are grateful to Ian A. Williams of the Department of Philology, University of Cantabria, for the language revision of the manuscript. We also acknowledge the excellent technical assistance of Marisa Junco, Saray Pereda, Raquel Garcia Ceballos, y Miguel Mier.

Abbreviations

ActRIIa	Activin receptor II a
Acvr1	Activin receptor I
Acvr2b	Activin receptor II b
Aldh1a2	Aldehyde dehydrogenase family 1 subfamily 2
AP	Antero-posterior
ASE	Asymmetric enhancer
Bmp	Bone morphogenetic protein
BPI	Bactericidal-permeability-increasing protein
Car	Caronte
Cer	Cerberus
CFC	Crypto-FRL1-cryptic
CHD	Congenital heart defects
Chrd	Chordin
CITED	CBP/p300 interacting transactivators with ED-rich termini
co-Smad	Common-partner Smad
CRELD1	Cysteine-rich protein with EGF-LIKE domain 1
Cx43	Connexin 43
Dand5	DAN domain family member 5
Dhh	Desert hedgehog
Disp	Dispatched
Dll1	Delta-like1
Dnal	Dynein axonemal intermediate chain type 1
Dnah7	Dynein heavy chain 7
Dnah9	Dynein heavy chain 9
Dnahc11	Dynein axonemal heavy chain 11
Dnahc2	Dynein axonemal heavy chain 2
Dnahc5	Dynein axonemal heavy chain 5
Dnai2	Dynein intermediate chain 2
Dnal1	Dynein axonemal light chain 1
DORV	Double-outlet right ventricle
DV	Dorso-ventral
Dync2h1	Dynein cytoplasmic 2 heavy chain 1
Dync2li1	Dynein cytoplasmic 2 light intermediate chain 1
E	Embryonic day

EGF	Epidermal growth factor
Fgf	Fibroblast growth factor 8
flh	Floating head
Fltn	Flectin
Fts	Fused toes
Gdf1	Growth differentiation factor 1
GJA1	Gap junction alpha-1 (Gap junction protein 43KD)
Hfh4	Hepatocyte nuclear factor forkhead homolog 4
HH	Hamburger Hamilton
hLAMP1	Heart-specific lectin-associated matrix protein-1
HNF3-β	Hepatocyte nuclear factor 3-β
ICS	Immotile-cilia syndrome
IFT	Intraflagellar transport
Ift172	Intraflagellar transport protein 172
Ift88	Intraflagellar transport protein 88
Ihh	Indian hedgehog
Inv	Inversion of embryonic turning
Invs	Inversin
IP6	Inositol hexakisphosphate
Ipk1	Inositol 1,3,4,5,6-pentakisphosphate 2-kinase
I-Smad	Inhibitory Smad
Iv	Inversus viscerum
Kif3a	Kinesin family member 3A
Kif3b	Kinesin family member 3B
Kif5C	Kinesin family member 5C
LEF1	Lymphoid enhancer binding factor 1
Lefty1	Left–right determination factor 1
Lefty2	Left–right determination factor 2
Lfng	Lunatic fringe
Lgl	Legless
LPM	Lateral plate mesoderm
LR	Left–right
Lrd	Left–right dynein
LSE	Left-side-specific enhancer
Mgat1	Mannoside acetylglucosaminyltransferase 1
Mid1	Midline 1
Mid2	Midline 2
NCX1	Sodium–calcium exchanger
Nkx3.2	Bagpipe homeobox gene 1 homolog
Nog	Noggin
NPHP2	Nefronophthisis type 2
Nt	No turning
Ntl	No tail

Ofd1	Oral facial digital syndrome 1
OMIM	Online Mendelian inheritance in man
PC1	Polycystin 1
PC2	Polycystin 2
PCD	Primary ciliary dyskinesia
Pcl2	Polycomblike 2
Pitx2	Paired-like homeodomain transcription factor 2
PKA	Protein kinase A
Pkd2	Polycystic kidney disease 2
PKI	Protein kinase A inhibitor
PKIα	PKA inhibitor
PLUNC	Palate lung nasal epithelium clone
Pol-λ	DNA polymerase X family member λ
Psck6	Proprotein convertase subtilisin/kexin type 6
Ptc	Patched
RA	Retinoic acid
R-Smad	Receptor-regulated Smad
Rttn	Rotatin
Shh	Sonic hedgehog
Slc8a1	Solute carrier family 8
Smo	Smoothened
SSCP	Single-strand conformation polymorphism
Stil	Scl/Tal1 interrupting locus
Sufu	Suppressor of fused
T	Brachyury
TCF	Transcription factor T cell specific
TFAP2	Transcription factor AP2
TGA	Transposition of the great arteries
TGFβ	Transforming growth factor β
THRAP2	Thyroid-hormone-receptor associated protein 2
TTTC8	Tetratricopeptide repeat domain 8
Wnt	Amalgam from wingless (wg) and mammary tumor integration (int)
Zic3	Zinc-finger protein of the cerebellum 3

1 Introduction

Externally the vertebrate body plan presents a bilateral symmetry in relation to the midline, with two lateral essentially symmetric halves. However, inside the body the distribution of the visceral organs follows a very particular pattern that is not symmetrical in relation to the midline. The asymmetries in organ arrangement are characteristic and fixed. Many organs are unpaired and accommodate on one side of the internal body cavities. The liver and vena cava are examples of organs located on the right while the spleen and the aortic arch are examples of organs located on the left. Paired organs, such as the lungs and the kidneys, are usually arranged more or less bilaterally but always display characteristic and constant asymmetries between the two sides. Besides their position, many organs also show clear morphologic asymmetries between the two sides of the organ itself, the best example probably being the heart. As is well known, this pattern of internal organ distribution is highly conserved not only within species but also across vertebrate species.

In vertebrates, specification of the left–right (LR) axis occurs in coordination with the anterior-posterior (AP) and dorso-ventral (DV) axes and involves a complex and highly regulated process that starts with the initial breaking of the bilateral symmetry of the embryo and concludes when the asymmetric morphogenesis of the organs has been completed.

Alterations in LR axis development can lead to variations in internal organ arrangement that may interfere to a greater or lesser extent with normal development. In recent years an enormous amount of information has emerged on the development of the LR axis. This has led to a better understanding of genetic, molecular, and cellular mechanisms regulating the establishment of body asymmetries. Our goal is to present a comprehensive, integrated review of current knowledge in the field based on vertebrate organisms, principally mammals and avian embryos, and with human references when appropriate. The nomenclature used for the genes in this review is that of Mouse Genome Informatics, with synonyms also provided when appropriate.

Because of the amplitude of the field it is not possible to cover all aspects in a single review. Some topics such as the complex and expanding field of cerebral asymmetries were felt to fall outside the scope of the review and have, therefore, not been included.

1.1 The First Morphological Asymmetries in the Left–Right Axis

During embryonic development the earliest manifestation of major morphological asymmetry is the looping of the heart that occurs on embryonic day 23 in humans, E8.5 (embryonic day 8.5) in the mouse and by stage 10–11 of Hamburger Hamilton (HH, (Hamburger and Hamilton 1951) in the chick (Fig. 1). The heart develops from two bilaterally symmetric primordia that form in the splanchnopleure, on either side of the embryo. Concomitantly with the folding of the embryo, these two primordia gradually converge at the ventral midline where they fuse, forming the linear heart tube (Icardo and Ojeda 1984). Invariably the tubular heart bends toward the right to form the cardiac loop, in a process that is highly conserved in vertebrates, independently of the variations in the final morphology of the adult heart (Icardo et al. 2004). This fact has been assumed to reflect a common evolutionary origin.

In the chick embryo a slight transient asymmetry is apparent in Hensen's node at gastrulation during stages 4 to 7HH (see Fig. 6B; Cooke 1995; Cooke 2004; Dathe et al. 2002; Hara 1978). This asymmetry consists in the greater prominence of the right than the left side of the node. This asymmetry is also manifested in the primitive streak in which the right lip is also more prominent that the left lip. It is important to mention that these morphological asymmetries arise before asymmetric gene expression is established in the node, but their significance is presently unknown.

The next manifestation of asymmetry in embryonic development is the axial rotation of the embryo, which is particularly interesting in the mouse embryo. During early embryogenesis, the rodent embryo presents a characteristic cup-shape with the endoderm (ventral) lining the convex outer aspect of the embryo and the ectoderm (dorsal) in the concavity (Fig. 2). This type of configuration is referred to as "inversion of the germ layers" and a suggestion has been made that it presents the evolutionary advantage of reducing the space necessary for implantation (Snell and Stevens 1966). At the 6–8-somite stage the mouse embryo begins process of "turning," involving a 180° anticlockwise twisting of the embryo along its AP axis (Beddington and Robertson 1999; Theiler 1989). After this turning the initial location of the germ layers has reversed, the endoderm lining the gut tube in the concavity and the ectoderm on the convex side forming the neural tube. As a consequence of this rotation, the embryo has changed its initial lordotic position to the fetal position; the vitelline vessels become located along the left side of the body and the tail directed towards the right side of the head. At present, the mechanism and molecular control of embryonic turning is poorly understood. This turning is a characteristic of rodent embryos and is not shared by other mammals; early rabbit and human embryos have a flat shape more similar to the chick embryo (Fischer et al. 2002; Muller and O'Rahilly 1987).

In stark contrast, the pregastrulation chick embryo is a flat bilayered disk that after gastrulation progressively rotates 90 °C clockwise around its AP axis, causing its right side to face upward and its left side to face down towards the vitellum.

1.2 Terminology

At present there is no consensus regarding the terminology for the position of the organs in relation to the LR axis and its deviations from normal. Therefore, we shall briefly define the terms used in this review, as shown in Table 1.

Table 1 Left–right terminology as used in this review

Term	Definition
Situs solitus	Normal arrangement and shape of internal organs
Situs inversus or Situs inversus totalis	Reversed arrangement of organs with respect to the midline Perfect mirror image of *situs solitus*
Situs ambiguus	Any other arrangement of organs other than *situs solitus* or *situs inversus*
Heterotaxia	Subtype of *situs ambiguus* No concordance in the *situs* of various organs
Isomerism or Sequence of isomerism	Subtype of *situs ambiguus* The two halves of an organ or the two asymmetric bilateral organs appear in a symmetric arrangement
Dextrocardia	Inverted position of the heart with the apex pointing toward the right
Mesocardia	Central position of the heart in the thorax

The term *situs solitus* indicates the normal condition, in which the organs have the usual morphology and arrangement, as described by gross anatomy in humans. The situation in which the shape and arrangement of the organs is a perfect mirror image of the *situs solitus*, is known as *situs inversus* or *situs inversus totalis* (Fig. 3). Any arrangement of organs other than *situs solitus* or *situs inversus* is referred to as *situs ambiguus. Situs ambiguus* is thus a broad term covering any deviation from *situs solitus* except for *situs inversus*, ranging from a mild defect in the position of an internal organ to severe phenotypic syndromes.

The main subtypes in *situs ambiguus* are *heterotaxia* and *isomerism. Heterotaxia* is the condition in which the *situs* of various organs is not concordant: that is, the position of some organs corresponds to *situs solitus* while others organs show different arrangements. *Heterotaxia* includes a wide range of defects, from mild to extremely severe, and may affect several organs. Examples of *heterotaxia* are intestinal malrotations, biliary tract anomalies, and a myriad of heart defects. *Isomerism* (also referred to as *sequence of isomerism*) is a condition in which the two sides of a body cavity or the two sides of an asymmetric organ appear in a symmetric arrangement. When the two sides resemble the normal left side, then the *isomerism* is termed left- or *levo-isomerism*. Conversely, when the two halves replicate the right side, the *isomerism* is called right- or *dextro-isomerism*. Examples of *isomerism* include symmetric lungs or atria and midline liver, all of them being severe conditions requiring complex surgical reconstructions and often incompatible with life. Recently, as we shall see later in this review, other pathologic conditions such as cystic renal disease have been shown to be associated with defects in LR patterning.

Another term broadly used in the field is "laterality defect," which designates, albeit imprecisely, any alteration in the *situs* of the organs.

The heart is the organ with the most pronounced and functionally significant asymmetries. It is also highly sensitive to perturbations in developmental pathways, which gives rise to a complex and rich pathology. Furthermore, the position of the heart is often used as the "readout" of the body *situs* mainly because it is the first organ to show LR asymmetry. All this had led to the creation of a heart-specific terminology. In the *situs solitus* the apex of the heart points to the left (*levocardia*). *Dextrocardia* is used when the apex points to the right and *mesocardia* refers to a heart centered in the thorax with the apex in the midline, possibly reflecting failure in looping.

1.3 The Asymmetries in the Left–Right Axis Are Highly Conserved

Almost all individuals in a vertebrate population display *situs solitus*. The thoracic cavity contains the lungs and the heart, the latter located in the mediastinum or central region of the chest, with the apex pointing to the left. Because of the asymmetric position of the heart, the right lung is larger than the left. In humans the right lung consists of three lobes while the left lung has only two. In mouse, this difference is even greater, as the right lung consists of four lobes whereas the left lung has only one. In the abdominal cavity the largest organ is the liver, which is situated in the right hypochondrium while the stomach and spleen lie in the left. The human liver has four lobes, one of which, the left lobe, crosses the midline to enter the left hypochondrium.

The *situs solitus* is a highly conserved feature not only in individuals within the same species but also across different species, with the particularities proper to each species. Studies performed in adult human populations have shown that variations in LR asymmetries are rare, with an incidence of 1:10,000 to 1:20,000 individuals (see for example Berg et al. 2003; Burn 1991; Fujinaga 1997; Lin et al. 2000; Skandalakis and Gray 1994).

The high degree of invariability in the LR organization of the internal organs is commonly assumed to reflect the benefits that this arrangement provides for their harmonious development and functioning. Indeed the organs are compactly packed inside the body cavities, the anatomy of each organ perfectly adapted to that of the surrounding organs and body walls. It is easy to assume that the evolution has selected this particular order as the one providing the best conditions for the coordinated functioning of the organs.

Since *situs inversus* is by definition a mirror image of the *situs solitus*, the two arrangements should offer the same benefits for the harmonic relationships and functioning of the organs. And indeed this is the case, since individuals with *situs inversus* carry no clinical manifestation of disease and have been identified by a chance finding on close clinical examination, a surgical procedure or at postmortem. The existence of individuals with complete inversions of the visceral *situs* is well documented and has been known for centuries (McMannus 2002). Why the *situs solitus* is always favored in Nature remains presently unknown.

On the other hand, an increased incidence of congenital heart defects (CHD) has been reported to occur in individuals with *situs inversus* (Ferencz et al. 1985; Nugent et al. 1994; Sternick et al. 2004) although other studies have failed to detect such increase in the incidence (Skandalakis and Gray 1994).

The fact that the asymmetries in the vertebrate body are highly invariant and well-conserved across species, together with the negative effects of uncontrolled laterality, is a strong indication that there is a robust developmental control. Indeed, a formidable research effort in recent years has begun to reveal the complex and intricate processes implied in this control (for recent reviews see Burdine and Schier 2000; Capdevila et al. 2000; Hamada et al. 2002; Levin 2005; Yost 2001). Given the consistency in the handedness of asymmetries between species, it was formerly assumed that their control mechanisms would be highly conserved. However, the discovery of species-specific mechanisms has suggested that different species may have developed different strategies that lead to a final common pattern of laterality. The differences may initially be marked by the variable early embryonic development of the species, followed by a convergence to a phylotypic LR stage in the process of establishing laterality (Yost 2001) with the different species later diverging to form the final morphology. This is similar to the Von Baer phylotypic stage in early embryonic development.

1.4 The Situs of Individual Organs Is Independently Established

One interesting question is whether the *situs* of an individual organ is specified independently or in relation to the *situs* of adjacent organs. Several experimental situations have shown that the laterality information for each organ is set independently, that of the neighbors disregarded (Chin et al. 2000; Levin et al. 1997). Therefore, each organ unit appears to receive and respond to the LR pathway independently. In the absence of signaling, each unit makes a random decision and the final outcome is the sum of independent LR decisions with respect to each other.

1.5 Left–Right Morphogenesis

The study of organ morphogenesis reveals that the early primordia of most organs start in a symmetric LR position, but then asymmetries are superimposed in different ways. There are three main types of process leading to the generation of organ asymmetries, these are: bending and rotation movements, asymmetric regression/persistence, and differential growth. In most cases, organ asymmetries are achieved through a combination of these three types of process (Carlson 2000; Moore and Persaud 1998; Sadler 2004).

Bending and rotation movements mostly affect tubular organs such as the heart and the gut. Both the heart and the gut acquire their asymmetries through a series of bending and rotation movements that are responsible for confronting appropriate parts of the organs for future morphogenesis. In the gut, this process also establishes the appropriate association between different organs and eventually locates each organ in its correct position. The morphogenesis of the heart is particularly complex and will be treated separately.

The morphogenesis of most parts of the digestive tract involves a process of rotation as in the stomach, the liver and the gut. For example, the primary intestinal loop rotates around an axis formed by the superior mesenteric artery. If observed ventrally, this rotation occurs counterclockwise through approximately 270°. The rotation of the gut is particularly observable in *Xenopus* since it can be seen through the transparent body wall. The spiral coiling of the gut in *Xenopus* usually serves as "readout" of the body *situs*. The stomach also undergoes a 90° clockwise rotation (viewed ventrally) so that its initial left face turns to be placed anteriorly. The liver also begins its development in the midline and will eventually adopt its final position and form through complex rotation and differential growth (Carlson 2000; Moore and Persaud 1998; Sadler 2004).

The origin of many asymmetric organs is bilateral and symmetric, but later some of the primordia partially or completely regress to leave unilateral or highly asymmetric organs. For example the spleen develops as two bilateral primordia but subsequently the right primordium disappears (Patterson et al. 2000). The development of the vascular system is also an excellent example of differential regression/persistence between bilateral components. The arterial system first arises bilaterally symmetric. As the pharyngeal arches progressively form, the aortic sac provides them with a branch, called the aortic arches, and five or six pairs of aortic arches develop which run through the branchial arches towards the dorsal aorta, which at that point is also bilateral and symmetric in its cranial part. During subsequent development this pattern is deeply modified, mainly by the regression of some components, to establish the well-known asymmetric pattern of the head and neck arteries (Davies and Guest 2003). This remodeling depends on the presence and integrity of the cardiac neural crest (Hutson and Kirby 2003). The first two pairs of aortic arches regress and the third evolves to form the paired common and internal carotid arteries. The left and right fourth aortic arches follow asymmetric development contributing to the definitive aortic arch on the left and the right subclavian artery on the right. The fifth pair is transitory and in some species does not even form. The sixth pair, also known as the pulmonary arch, contributes to the proximal segment of the right pulmonary artery on the right and to the left pulmonary artery and ductus arteriosus on the left (Carlson 2000; Moore and Persaud 1998; Sadler 2004).

As with many aspects of heart morphogenesis, the final morphogenesis of the arterial system in different species of vertebrates present peculiarities that are outside of the scope of this review. It is, however, of interest to mention that the definitive aortic arch in birds is right-sided and derives in part from the right fourth aortic arch (Romanoff 1960). At present, no explanation has been given as to why the directionality of this marked asymmetry is inverted between birds and mammals.

The venous system, like the arterial system, develops from a symmetric origin with the formation of three pairs of major veins, the cardinal, vitelline, and umbilical veins. Formation of the definitive venous pattern is a complex process that involves obliteration of several portions and generation of important anastomoses (Ruscazio et al. 1998). The complex development of both the arterial and venous systems explains the relatively frequent observation of deviated patterns that may have clinical significance.

The third mechanism used by the embryo to create asymmetries from bilateral primordia is differential growth. It is typical of branching organs such as the liver and the lungs. The lung bud arises as a ventral outgrowth from the ventral wall of the foregut. As the bud grows caudally, it begins to divide into two branches forming the bronchial buds that will develop into the main bronchi. From this point LR asymmetry is clear since the right bronchi always show enhanced branching compared to the left and, therefore, give off more secondary bronchi than the left branch. Branching is regulated by epithelial-mesenchymal interactions between the endoderm and mesoderm.

1.5.1 Vertebrate Heart Development

The heart is a complex organ, the morphogenesis of which involves folding, rotation, differential growth, and asymmetric regression/persistence. Cardiac precursors are specified in the epiblast of the gastrula embryo quite early during development (Antin et al. 1994). After migration through the primitive streak, the heart precursors arrange into paired (primary) heart fields that form a crescent of mesoderm in the most anterior part of the embryo (Icardo 1996). The mesodermal crescent contain myocardial and endocardial precursors; the latter organize into endocardial tubes wrapped by myocardium. Concomitantly with the formation of the body walls, the crescents converge to fuse in the ventral midline to form the primitive heart tube. The primitive heart tube is already regionalized and the presence of transversal grooves divides it into the caudal inflow region and the cranial ventricular area that continues cranially with the outflow region.

The formation of the heart is completed by the addition of another group of cells at the growing arterial pole. This group constitutes the so-called secondary or anterior heart field, whose extension varies according to different authors (Franco and Icardo 2001; Kelly et al. 2001; Waldo et al. 2005a; Waldo et al. 2005b; Waldo et al. 2001). The secondary heart field is formed by cells from the ventral pharynx and their incorporation into the heart requires signaling by the neural crest cells that migrate through the branchial arches to colonize the outflow track and to a lesser extent the inflow track (Hutson and Kirby 2003).

Since cardiac fate is specified very early, during gastrulation, it is likely that the cardiac precursors are exposed to left–right asymmetric signals. Also, the asymmetric gene expressions established in the lateral plate mesoderm (LPM) after gastrulation (see Sect. 2.3) reach the caudal part of the primary heart field and later expressions also affect derivatives of the secondary heart field. Therefore the right and left primary heart fields presumably have a different history of molecular influences that make them non-equivalent. Similarly, owing to differential gene expression, is it likely that each side of the primary and secondary heart fields is heterogeneous along the AP axis (Ramsdell 2005). In this regard, it is of interest to mention that some extracellular proteins have been shown to be asymmetrically expressed in the chick primary heart fields. These are flectin and hLAMP1 (Smith et al. 1997; Tsuda et al. 1996), which are expressed on the left, and fibrillin2, which shows stronger expression on the right (Table 2; Smith et al. 1997). Flectin also exhibits a marked asymmetric expression at later stages of mouse heart development (Table 6; Tsuda et al. 1998).

As soon as the heart tube forms, it starts to bend and forms the cardiac loop that is invariably convex to the right (Icardo 1996). As mentioned previously, this is a highly conserved process in vertebrates that occurs quite early in embryonic development and is considered to be the first evidence of morphological LR asymmetry. The looping process is quite complex involving ventral bending and rightward rotation and has been extensively studied (Icardo 1996; Icardo 1997; Manner 2000; Manner 2004). The resulting "C" shaped cardiac loop presents a rearrangement of the spatial location of the different chambers for proper connection and septation to occur. The looping continues into an "S" shaped heart, which places the ventricles ventrally in relation to the atria.

An important consideration about the heart is that the right and left ventricles are specified along the AP axis of the tubular heart, a phenomenon sometimes referred to as "in series" development. Both ventricular chambers adopt the LR juxtaposition as a result of cardiac looping. However, this is not the case with the atria: these develop from a common progenitor, the common atrium, which later divides into the left and right atrial chambers as a result of the growth of the interatrial septum. This is sometimes referred to as "parallel" development. As the two atria become individualized their walls specialize and they establish different vascular connections. From its earliest development, the common atrium receives asymmetric LR signaling that is thought to direct the different development of the two sides. Failure in this early specification process lies at the root of atrial isomerism. The interatrial septum originates from cells with left-sided gene expression (Franco and Campione 2003) and is thought to require left-side information for its formation. If both sides of the common atrium receive left-sided signals, then the two atria develop as left atria resulting in left *isomerism* and vice versa. In support of this view is the observation of severe defects of the interatrial septum in individuals with right atrial *isomerism* (Bowers et al. 1996). At the morphological level no case of ventricular *isomerism* has been reported, probably due to the particular in-series development of the two ventricles. However, there is some evidence indicating the existence of molecular isomerism in the ventricular region, correlating with the development of double outcome of right ventricle (DORV; Campione et al. 2001).

Table 2 Gene with asymmetric LR expression during the early development of the chick embryo

Gene symbol	Name; role	Asymmetric expression	Stage	Reference
Activin-βB	TGF-*β*-family signaling molecule	Right side of Hensen's node	3HH to 5^+HH	Levin et al. 1997
ActRIIa	Activin receptor II a	Right side of primitive streak and Hensen's node	4 and 4^+HH	Levin et al. 1995
Bmp4	BMP family signaling molecule	Right side of Hensen's node	4 to 8HH	Monsoro-Burq and Le Douarin 2000
Caronte	DAN domain family member; BMP and Nodal antagonist	Left paraxial mesoderm and left lateral plate mesoderm	7HH to 9HH	Yokouchi et al. 1999; Zhu et al. 1999; Rodriguez Esteban et al. 1999
Cited2	Transcriptional repressor of HIF1; nuclear transactivator	Right side of the embryo	During 5HH	Schlange et al. 2000
Cx43	Gap junction protein	Right side of Hensen's node	During 5^-HH	Levin and Mercola 1999
Dll1	Delta-like1; DSL ligand of Notch pathway	Left side of Hensen's node	During 5HH	Raya et al. 2004
Fgf8	Fibroblast growth factor 8	Right side of the posterior Hensen's node	6 to 8HH	Boettger et al. 1999
Fibrillin-related	Extracellular matrix molecule	Stronger on the right precardiac field (immunohistochemistry)	5 to 7HH	Smith et al. 1997; Wunsch et al. 1994
Flectin	Extracellular matrix molecule	Stronger on the left precardiac mesoderm (immunohistochemistry)	$7^+/8^-$HH	Tsuda et al. 1996
Follistatin	Activin antagonist	Right of Hensen's node	4HH	Levin 1998b

Table 2 (continued)

Gene symbol	Name; role	Asymmetric expression	Stage	Reference
Gli1	GLI-Kruppel family member; mediator of Shh signaling	Left side of Hensen's node (ectoderm layer)	5 to 7HH	Granata and Quaderi 2005
Gli2	GLI-Kruppel family member; mediator of Shh signaling	Left side of Hensen's node (ectoderm layer)	5 to 7HH	Granata and Quaderi 2005
Gli3	GLI-Kruppel family member; mediator of Shh signaling	Right side of Hensen's node (mesoderm layer)	6 to 7HH	Granata and Quaderi 2005
hLAMP1	Heart-specific lectin-associated matrix protein-1	Stronger on the left precardiac and ventral foregut regions (Immunohistochemistry)	6^- to 8HH	Smith et al. 1997
HNF3β	Hepatocyte nuclear factor 3-*β*	Small part of the left side of the primitive ridge, just posterior to the node	During 4^-HH	Levin et al. 1995
Islet-1	LIM homeobox gene 1	Stronger on the left rostrolateral mesoderm and underlying endoderm (cardiac progenitors)	4^+HH to 7^+HH	Yuan and Schoenwolf 2000
Kif5C	Kinesin family member 5C	Stronger on the right side of Hensen's node	4 to 7HH	Dathe et al. 2004
Lefty	Left–right determination factor 2; TGF-*β* family	Left side of the node, prospective floorplate, and precordal notochord	4 to 8HH	Ishimaru et al. 2000; Rodriguez-Esteban et al. 1999
		Posterior left LPM	8 to 10HH	

Table 2 (continued)

Gene symbol	Name; role	Asymmetric expression	Stage	Reference
Mid1	Microtubule-associated ubiquitin ligase	Right side of Hensen's node	5 to 7HH	Granata and Quaderi 2003
Mid2	Microtubule-associated	Right side of Hensen's node	5 to 7HH	Granata et al. 2005
N-Cadherin	Adhesion molecule	Stronger on the left side of the primitive streak	3^+ to 5HH	Garcia-Castro et al. 2000
		Right side of Hensen's node	4^+ to 5HH	
NCX1	Sodium-calcium exchanter	Right side of Hensen's node	4 to 5HH	Linask et al. 2001
Nkx3.2	Bagpipe homeobox gene 1 homolog; transcription factor	Anterior left LPM	10 to 13HH	Schneider et al. 1999
Nodal	TGF-β family signaling molecule	Cells lateral and anterior to the left side of Hensen's node (later bilateral 9HH)	6 to 9HH	Levin et al. 1995
		Left LPM	7 to 11HH	
Patched	Shh receptor	Lateral to the left side of Hensen's node	5^+ to 7HH	Pagan-Westphal and Tabin 1998
Pcl2	Polycomb-like 2; transcriptional repressor	Right side of Hensen's node	5 to 8HH	Wang et al. 2004
Pitx2	Paired-like homeodomain transcription factor	Left LPM and derivatives	From 7^+HH	Logan et al. 1998; Ryan et al. 1998; Piedra et al. 1998
PKIα	PKA inhibitor	Stronger on the right side of the Hensen's node	6 to 7^+HH	Kawakami and Nakanishi 2001; Rodriguez-Esteban et al. 2001

Table 2 (continued)

Gene symbol	Name; role	Asymmetric expression	Stage	Reference
Shh	Sonic hedgehog; signaling molecule	Left side of Hensen's node	5 to 7HH	Levin et al. 1995
Snail1	Transcription factor	Right posterior cardiogenic LPM	8 to 10HH	Isaac et al. 1997
Wnt8c	Wingless-related MMTV integration site 8C	Right side of Hensen's node	During 5HH	Levin 1998a; Rodriguez-Esteban et al. 2001

Regarding LR asymmetries, there are two major phenotypes that affect cardiac looping. Some phenotypes show randomization of cardiac looping. This indicates that the genes that cause randomization are involved not primarily in the process of looping itself but in the selection of the direction of looping. In contrast a second type of phenotype shows *mesocardia* or impossibility for cardiac looping. The genes causing this second phenotype are probably involved in the break of bilateral symmetry or cardiac morphogenesis.

It has been suggested that impaired LR signaling of the heart is a putative molecular determinant of common cardiac congenital malformations (Franco and Campione 2003).

2
Establishment of Left–Right Asymmetry

In the mid-1990s, molecular analysis performed in early embryos of several species demonstrated that a number of genes had a marked LR asymmetry in their pattern of expression, well before the start of morphological asymmetries, namely heart looping. These observations led to the idea that the LR axis is established during gastrulation, much earlier than previously thought, and they stimulated the search for the process that would trigger and control LR asymmetries.

Theoretically, at least from an academic point of view, the process of LR specification can be considered to evolve in different phases (Capdevila et al. 2000; Hamada et al. 2002). The initial step involves the specification of the two halves of the embryo, left and right, as different. Depending on the species, this process either occurs in the node, or equivalent structure, or converges to it, so that the node acquires the information and asymmetric gene expression. Then the information has to be transferred from the node to the LPM where it is apparent as side-specific domains of gene expression. Finally this information has to be translated into the development of the actual asymmetrical morphogenesis of the organs.

2.1
Breaking of the Initial Symmetry in the Embryo

The LR axis is determined in coordination with the DV and AP axes of the embryo. This implies that the LR axis is predetermined by the other two axes, except for its polarity, which has to be selected, and is not random. As mentioned previously, most species exhibit an almost absolute bias towards the *situs solitus* body plan, while the reversed mirror image of *situs inversus* is very seldom found in Nature. Thus, the initial step in LR patterning includes the breaking of the initial LR symmetry of the embryo and the implementation of polarity in the axis. Several observations indicate that these two concomitant events are independently controlled, and so each of them may fail independently. When breaking of symmetry fails, the subsequently development of the organs is arrested or is highly aberrant. When selection of polarity fails, this is randomly applied and leads to a population with 50% individuals displaying *situs solitus* and 50% *situs inversus*. Our current knowledge indicates that inconsistent decisions regarding establishment of LR handedness may lead to severe alteration of visceral development or even embryo death. Therefore, it is of the greatest interest to investigate and understand how the breaking of symmetry occurs and how the choice of handed asymmetry is made and maintained.

Several theoretical models based on experimental data have attempted to explain how these early processes are achieved. We shall consider each of the models in turn.

2.1.1
The Nodal-Flow Model

A few years ago, the study of the kinesin molecular motors involved in ciliogenesis (see Appendix) led Hirokawa et al. (2006) to concentrate on the morphology and function of the monocilia (or primary cilia) present in the mouse node (Fig. 4) (Nonaka et al. 1998; Okada et al. 1999; Takeda et al. 1999). Targeted disruption of *Kif3a*[1] and *Kif3b*, two components of the heterotrimeric kinesin II required for ciliary assembly, revealed the association between absence of node monocilia and randomization of cardiac looping (Marszalek et al. 1999; Nonaka et al. 1998; see Appendix). Close anatomical examinations showed that the mouse node was composed of two layers of cells, one ventral and one dorsal (Fig. 4; Bellomo et al. 1996; Sulik et al. 1994). The cells in the ventral layer are columnar, exhibit a prominent single central cilium and give rise to the notochord. The dorsal layer is formed by columnar cells that will form the floor plate in the ventral neural tube (Beddington 1994). As occurs with many monocilia in embryonic cells, the cilium of the ventral nodal cells had a 9+0 axonemal arrangement, an organization considered typical of immotile cilia (see Appendix). However, Hirokawa and his colleagues demonstrated by video microscopy that these cilia did move in a particular clockwise vortical movement (Nonaka et al. 1998). Furthermore, they showed that the combined movement of all the cilia in the node produced a neat and constant leftward flow of the node fluid. From these results they elaborated a new, original, and elegant model, the "nodal-flow model," to explain the initiation of LR asymmetries in the mouse embryo (Fig. 5). The model postulates that a putative morphogen, symmetrically produced and secreted into the nodal fluid, becomes displaced to the left by the flow produced by the cilia. The accumulation on the left of this putative morphogen would discriminate between the two sides of the embryo and start LR asymmetries (Fig. 5; Nonaka et al. 1998; Okada et al. 1999; Takeda et al. 1999).

Although several candidates have been proposed, the existence and nature of this putative morphogen remains to be elucidated. In this regard, the same group

[1] Throughout the manuscript we have adopted the Mouse Genetic Informatics gene nomenclature. The first time a gene is named its symbol is indicated in parentheses, as well as synonyms when appropriate. Gene names and symbols are italicized but corresponding protein names are not italicized. Human gene names and symbols are uppercase.

of investigators has recently identified huge membrane-sheathed objects named "nodal vesicular parcels" (NVPs), which are secreted and transported to the left by the nodal flow and constitute a brand new mode of extracellular transport of morphogens (Tanaka et al. 2005). Interestingly, signaling by Fibroblast growth factor (Fgf) triggers the secretion of the NVPs that carry Sonic hedgehog (Shh) and retinoic acid (RA). During their flow to the left the NVPs finally fragment into small particles that facilitate their association with the cell surface and therefore signaling. The cilia appear to facilitate the fragmentation of the NVPs.

The nodal-flow hypothesis considers that the nodal flow generated by motile cilia is the earliest symmetry-breaking event in mammalian embryos. According to this model, the initial breaking of symmetry takes place in the node and it is the leftward flow that imposes the handedness. As mentioned above, when breaking of symmetry fails, LR asymmetry still occurs but is random. The model predicts that absence of the nodal flow will lead to randomization of heart looping and organ *situs*, whereas a reversal in the direction of the flow would result in reversed visceral *situs*. Accordingly, the *Kif3a* and *Kif3b* mutants that lack cilia (Marszalek et al. 1999; Nonaka et al. 1998; Takeda et al. 1999), and also the mouse spontaneous mutation *inversus viscerum* (*iv*), which is caused by a mutation in the specific axonemal dynein *Dnahc11* (*Dynein axonemal heavy chain 11*, formerly called *left right dynein*, *lrd*; Capdevila et al. 2000) that renders the nodal cilia immotile, show randomization of *situs* (Brueckner et al. 1989; Hummel and Chapman 1959; Layton 1976; McGrath et al. 1992). Similarly, highly sophisticated experiments designed to modify the direction of nodal flow in cultured mouse embryos showed that the reversal of nodal flow was sufficient to reverse visceral *situs* (Nonaka et al. 2002). All the above-mentioned data strongly support the nodal-flow model. So far no spontaneous or induced mutants with right-sided nodal flow have been reported. Surprisingly, the only mouse mutation that causes 100% situs *inversus*, called *inversion of embryonic turning* (*inv*), has a slow, turbulent nodal flow directed to the left, instead of the rightward flow predicted by the model (Okada et al. 1999; Yokoyama et al. 1993).

Moreover, the nodal-flow model has the added interest of providing a link between a physical structure, the cilia, and the handedness of the organ *situs*, something postulated by another model, the "chiral molecule model", put forward by Brown and Wolpert (1990). This model postulates that a chiral molecule, fixed in a particular orientation to the AP and DV axes, could mediate the asymmetric distribution of other components and therefore provide the initial specification of handedness. It could be that the nodal cilia or some of the ciliary components, with the particular orientation of the cilia and the specific architecture of the node, constitute the presumed chiral molecule, or molecular complex, proposed by Brown and Wolpert as responsible for the initiation of LR asymmetry.

The almost absolute invariability in the handedness of the organ *situs* among species has been taken as an indication of a common underlying mechanism. Therefore, it is of interest to determine whether this nodal-flow model applies in species other than mice. Monocilia with rotational movement have been observed in another mammal, the rabbit, creating a left-side flow in the notochordal plate, considered the equivalent of the ventral node in the mouse (Okada et al. 2005). However, the movement of the cilia may appear later than in the mouse, after the onset of asymmetric gene expression (Fischer et al. 2002; Okada et al. 2005; Tabin 2005). Similarly the nodal-flow has also been demonstrated in two species of fish, zebrafish and medakafish, very useful as anamniote vertebrate models. The monocilia in the zebrafish and medakafish organizer, the Kupffer's vesicle, have been shown to behave as in mice: they are motile and generate a left directional fluid flow just prior to the onset of asymmetric gene expression in lateral cells. If the generation of this flow is experimentally prevented, the development of the LR axis is consequently altered (Essner et al. 2005; Kawakami et al. 2005; Okada et al. 2005).

Monocilia have also been observed in Hensen's node in chicken embryos (Fig. 6) and the equivalent structure in *Xenopus* (dorsal blastopore; Essner et al. 2002). However, in these two species asymmetric gene expression has been detected before the appearance of the cilia, indicating that the cilia might not be responsible for initiation of asymmetries (Levin 2005). The architecture of Hensen's node in the chick also makes it difficult to envisage a way to apply the nodal-flow hypothesis (Fig. 6). Hensen's node cells connect the epiblast with the endoderm without the formation of a clear space or concavity in which the cilia could create nodal flow. The monocilia in the chick node are relatively far smaller than in the mouse (Fig. 6C) and even if they were motile, which is presently unknown, it is unclear how a nodal flow can be established that would be similar or equivalent to that in the mouse. The node does form a depression on its dorsal aspect but monocilia have been described on the ventral side of the node (Essner et al. 2002; Manner 2001). And, as mentioned, even if nodal flow were demonstrated, it would not be the first asymmetric manifestation of the embryo, since several asymmetric gene expressions have been detected earlier (Levin et al. 2002).

In summary, it is unclear whether the nodal-flow model can be generalized to most species. However, it could be that the initiation of LR asymmetries follows different strategies depending on the species, as has been shown for the AP and DV axes.

There are an increasing number of mutations affecting ciliary morphogenesis or function that also result in modification of organ *situs*, both in human and mice (see Sects. 4 and 5). This fact indicates a strong link between dysfunctional cilia and the establishment of LR asymmetry. In human pathology, a group of syndromes linking ciliary alterations with *situs inversus* is known to exist since Afzelius (Afzelius 1976; Afzelius 1979). He showed that the triad of symptoms present in Kartagener syndrome—bronchiectasis, sinusitis, and *situs inversus*—are caused by dysfunctional cilia due to the absence of the outer dynein arms (see Appendix). Kartagener syndrome (OMIM 244400) is part of the heterogeneous primary ciliary dyskinesia (PCD) disease (also formerly called immotile-cilia syndrome, ICS) the etiology of which resides in malformed cilia and flagellae (Afzelius 1976; see Sect. 5.1). We will refer extensively to the link between cilia and LR asymmetry in the course of this review although the details of the link are not completely known.

2.1.2 The Two-Cilia Model

One of the main criticisms of the nodal-flow model is its failure to provide a satisfactory explanation for the observation that embryos with immotile nodal cilia and those with total absence of nodal cilia have dissimilar alterations in LR patterning (Brueckner 2001; Tabin 2005; Tabin and Vogan 2003; Vogan and Tabin 1999; Wagner and Yost 2000). Mutations in axonemal dyneins such as *Dnahc11* (lrd), and in the forkhead transcription factor Foxj1, that result in paralyzed cilia in the node, are associated with a complete spectrum of patterns of left-side markers' expression in the LPM (namely bilateral, absent, left-sided, or right-sided) and subsequent laterality phenotypes. In contrast, mutations in genes that result in total absence of cilia in the node such as those required for ciliary assembly, including *Kif3a* and *Kif3b* (see Appendix), result in absent or bilateral expression of left-side markers in the LPM and a reduced range of laterality defects, *situs inversus* and left *isomerism* (Marszalek et al. 1999; Murcia et al. 2000; Nonaka et al. 1998; Takeda et al. 1999). In neither of these situations—immotile cilia or absence of cilia—is nodal flow generated and the model, therefore, predicts an identical outcome. The recent suggestion that the mere presence of cilia in the node may facilitate the breaking of the NVPs and therefore morphogen signaling, may provide an explanation for these differences (Hirokawa et al. 2006; Tanaka et al. 2005).

Interestingly McGrath et al. (2003) observed that there were two populations of monocilia in the mouse node. The cells located centrally in the node show motile monocilia that express the axonemal dynein *Dnahc11* whereas the cells in the periphery of the node have immotile monocilia that do not express *Dnahc11*. It is significant that all cilia in the node, including the immotile peripheral cilia, express the cation channel polycystin2 (PC2). PC2, which is encoded by the *polycystic kidney disease 2* (*Pkd2*) gene, is a calcium-release channel permeable to divalent cations. PC2 protein localizes within the primary cilia of the renal epithelium, where it has been shown to assemble with PC1 (Polycystin1) and cause intracellular calcium to increase in response to the bending of cilia under the shear stress of fluid flow in renal tubules (Nauli et al. 2003; Praetorius and Spring 2001). These findings lead to the proposal of the "two-cilia" model (McGrath and Brueckner 2003; McGrath et al. 2003; Tabin and Vogan 2003). This new model postulates a mechanosensory function for the nodal cilia, which is analogous to the function of monocilia in kidney cells (Praetorius and Spring 2001) and in which stimulated cilia trigger a calcium influx that can propagate to the surrounding cells through gap junctions. Thus, the central motile dynein-containing cilia produce the leftward flow of fluid whereas the peripheral immotile cilia act as mechanosensors to translate this flow into an intracellular influx of calcium. Due to the direction of the nodal flow, the cilia that become stimulated are those on the left periphery of the node (Fig. 5E). Accordingly, asymmetric calcium signaling appears at the left margin of the node coincident with the establishment of the nodal flow (McGrath et al. 2003). Furthermore, perinodal calcium signal is random in *Dnahc11* mutants and absent in *Pkd2* mutants, confirming that nodal-flow and PC2 are necessary for generating the increase in intracellular calcium in the left margin of the node (McGrath et al. 2003).

Interestingly, it has been shown that inositol 1,3,4,5,6-pentakisphosphate 2-kinase (Ipk1) is necessary for normal LR determination in zebrafish. Ipk1 is an enzyme responsible for generation of inositol hexakisphosphate (IP6). Loss of *ipk1* function randomizes molecular and morphological asymmetries and eliminates the left-biased Ca^{2+} flux in cells of Kupffer's vesicle, the zebrafish equivalent of the node (Sarmah et al. 2005). This permits one to envisage a situation in which once the Ca^{2+} influx occurs in the margin of the node, it is propagated to the left by an IP6-based activity (Sarmah et al. 2005).

It is worth noting that this hypothesis does not require the existence of a putative morphogen and that the nodal cilia are responsible both for generating and for receiving the stimulus that initiates LR asymmetry (Fig. 5D,E).

The expression of *Pkd2* in node and kidney monocilia provides an explanation for the observed link between the development of LR asymmetry and polycystic kidney disease. Loss of *Pkd2* function produces polycystic kidney disease and right isomerism (Pennekamp et al. 2002). Mutations in *inversin*, the gene mutated in *inv*, and in some components of the machinery of intraflagellar transport also combine laterality defects with polycystic kidney disease (McGrath and Brueckner 2003).

Like the nodal-flow model, the two-cilia model considers that the initial breaking of asymmetry occurs in the node as a consequence of the activity of the monocilia. It adds further insights into the way this flow is translated at cell level. The two-cilia model does provide an explanation for the differences in phenotype observed in the two mutants, those with immotile cilia and those lacking cilia. Mice with paralyzed nodal cilia are unable to set up the nodal flow but still retain the mechanosensory properties of the cilia able to detect subtle fluid movements (McGrath and Brueckner 2003; Tabin and Vogan 2003). Consistent with this proposal, mice unable to sense the nodal flow, for instance because of lack of polycystin, should exhibit the same phenotype as in complete absence of cilia. In fact, the phenotype observed in the absence of *Pkd2* is similar to that produced by total absence of nodal cilia (McGrath and Brueckner 2003; Pennekamp et al. 2002) supporting polycystin function as mechanotransducer of nodal flow.

Finally, it is interesting that experimental removal of the node in mouse embryos during late gastrulation results in embryos with abnormal LR patterning although the AP and DV axes show normal development. Such experiments underscore the importance of the node in generating or propagating left–right patterning information (Davidson et al. 1999).

2.1.3
The Gap-Junction Model

Another significant challenge to (criticism of) the hypothesis that the activity of the monocilia in the embryonic node is the first asymmetry that arises in the embryo comes from the study of other vertebrates, such as frogs and chicks. In these two animal models, asymmetric gene expression occurs long before the appearance of the node (Levin 2005) clearly demonstrating that the break in symmetry takes place earlier than formation of the nodal cilia.

In birds and frogs the cells of the early embryo are electrically coupled through gap junctions formed by connexins. In the early *Xenopus* embryo there is a dorso-ventral gradient of gap junctions; gap junctions are more abundant in dorsal blastomeres whereas ventral blastomeres are relatively isolated due to inactivity of rather than low number of gap junctions. There is also an area of junctional insulation on the ventral midline of the embryo (Guthrie et al. 1988; Levin 1998a; Levin 1998b; Levin and Mercola 1998a; Levin and Mercola 1998b). Experimental inversion of the dorso-ventral gap-junction gradient in the early *Xenopus* embryo, achieved through overexpression of dominant negative connexin in the dorsal blastomeres or wild type connexin in the ventral blastomeres, leads to randomization of organ *situs* (Levin and Mercola 1998b). Other experiments aimed at blocking gap-junction function also resulted in perturbations of the cascade of asymmetric gene activity and abnormal organ *situs*, indicating that the endogenous path of intercellular communication by gap junctions is involved in establishing correct LR patterning.

Gap junctions are also involved in the establishment of the left–right asymmetry in the chick (Levin and Mercola 1999). Experiments aimed at disrupting the endogenous gap junction path, either by removing tissue from the early chick blastoderm or experimentally prolonging the primitive streak, showed that circumferential signaling between the left and right halves of the epiblast is necessary for establishing correct LR axis. Genetic and pharmacological disruption of gap-junction communication in the chick gastrula also resulted in randomization of asymmetric gene expressions and organ situs (Levin and Mercola 1999).

All these experiments clearly indicate that gap-junction communication plays a critical role in LR patterning (Levin and Mercola 1998b; Levin and Mercola 1999). And this led to the proposal of a model in which one or several small molecule determinants travel through the circumferential pattern of gap-junction communications to accumulate on one side of the midline and thereby initiate LR asymmetry. It is important to emphasize that the function of gap junctions in LR patterning is upstream of asymmetric gene expression and prior to either node or blastopore formation. Serotonin has recently been proposed as an ideal candidate to travel across the gap junctions (Fukumoto et al. 2005).

Passage through gap junctions is bidirectional but the model requires a net unidirectional flow to allow accumulation of a determinant on one side of the midline zone of insulation. How this was accomplished was solved by identifying asymmetric function in the early *Xenopus* and chick embryo of molecules capable of creating voltage differences. Voltage differences provide an electrophoretic force for unidirectional flux of charged ions, suggesting the hypothesis that ion fluxes may be at the root of LR patterning in *Xenopus* and the chick (Levin et al. 2002).

In *Xenopus* asymmetric localization of maternal mRNA for the H^+/K^+-ATPase transporter between the left and the right side is visible at the 4-cell stage, demonstrating that LR asymmetry starts a few hours after fertilization (Levin et al. 2002). In chick, asymmetry in expression of the H^+/K^+-ATPase pump is not detected, but asymmetry in function is detected since there are differences in cell-membrane potential across the primitive streak that can be neutralized by inhibitors of the pump (Levin et al. 2002). A fusicoccin receptor, designated as 14–3-3E protein, is also asymmetrically expressed in the right but not the left blastomeres at the 2–4 cell stage (Bunney et al. 2003). Treatment with drugs modifying the K^+ flux or the H^+/K^+-ATPase function, or blocking 14–3-3E protein, affect LR development.

Therefore, the earliest known participant in specifying LR asymmetry in *Xenopus* and chick is the activity of the H^+/K^+-ATPase transporter. However, presently there is no evidence indicating a function for H^+/K^+-ATPase in laterality in species other than chick and *Xenopus*. The differential expression is translated into differences in membrane potential between the left and right sides of the primitive streak that could generate unidirectional flow through the gap junctions leading to the proposed accumulation of a putative LR determinant on the left side of the embryo. In the chick, depolarization caused by the H^+/K^+-ATPase activity on the left results in a transient rise of extracellular Ca^{2+} which preferentially activates *Notch* signaling, leading in turn to activation of *Nodal* expression (the left-side determinant, see Sects. 2.2 and 3.2; Raya and Belmonte 2006; Raya et al. 2004).

This model fits very well with experiments performed in chick and *Xenopus* that showed the importance of the peripheral tissues in conveying the information on LR asymmetry to the node (Hyatt and Yost 1998; Levin and Mercola 1999; Pagan-Westphal and Tabin 1998). In chick, when the LR axis of Hensen's node is experimentally inverted by a rotation of 180°, the peripheral tissues inductively interact with the node to re-specify its LR information according to that of the rest of the embryo. Beyond a certain stage the LR information on the node is not labile and cannot be re-specified if rotated (Pagan-Westphal and Tabin 1998).

It is worth mentioning that the implication of gap-junction communication in establishing LR asymmetries is strongly supported by studies in human indicating that some cases of heterotaxia arise from mutations in *GJA1* (*gap junction alpha1*, also *CONNEXIN43*; Britz-Cunningham et al. 1995; see Sect. 5). Since GJA1 is a structural component of gap junctions, the heterotaxia is assumed to be the result of disruption of gap-junction communication in the early stages of embryonic development.

2.2 Asymmetric Gene Expression in the Node

Avian LR axis specification involves the establishment of asymmetric gene expressions in Hensen's node (Table 2). Most of the genes expressed asymmetrically in Hensen's node are potent signaling molecules that establish hierarchies to set up discrete signaling pathways operating in the left and in the right side of the embryo.

The first report on asymmetric gene expression in the node came in 1995 when Cliff Tabin and his colleagues, in a seminal study, described several genes with asymmetric expression in Hensen's node (Levin et al. 1995; Fig. 7). Their study made it clear that there were obvious asymmetries in gene expression well before the onset of morphological asymmetries. Asymmetric expressions in the node began at stage 4HH, coincident with its formation, and persisted as the node regressed and the notochord started to form during the early somite stages. Through different experiments of gain and loss of function, these authors identified the hierarchies governing the relationships between these asymmetrically expressed signals. The pioneering work of Levin and colleagues (Levin et al. 1995) set the stage for numerous later studies that have exhaustively analyzed gene expression and function in relation to the chick node and have considerably extended our knowledge of the genetic control of LR asymmetries (Boettger et al. 1999; Garcia-Castro et al. 2000; Granata and Quaderi 2003; Granata and Quaderi 2005; Kawakami and Nakanishi 2001; Levin 1997; Levin et al. 1997; Monsoro-Burq and Le Douarin 2000; Monsoro-Burq and Le Douarin 2001; Raya et al. 2004; Rodriguez-Esteban et al. 2001; Shamim and Mason 1999; Wang et al. 2004).

Current understanding of gene expression and signaling interactions in the chick node is remarkably complex. A working diagram is provided in Fig. 7. Temporally, the first identified asymmetric expression detected in the chick embryo is the expression of *ActivinβB* on the right side of the node at stage 4HH (Levin et al. 1995). Activin signaling, acting through activin receptor IIa (AcvrIIa; see Sect. 3.3), induces the expression of *Bone morphogenetic protein 4* (*Bmp4*) in this side of the node (Monsoro-Burq and Le Douarin 2000; Monsoro-Burq and Le Douarin 2001).

Sonic hedgehog (*Shh*; see Sect. 3.1) is initially expressed symmetrically on both sides of the node but at stage 5HH its expression becomes asymmetric because it is suppressed from the right side by Bmp4 signaling. In the chick node, expression of *Shh* and *Bmp4* are complementary and mutually exclusive (Fig. 8; Monsoro-Burq and Le Douarin 2001; Piedra and Ros 2002). This mutual antagonism is mediated by *Midline1* (*Mid1*), a microtubule-ligated ubiquitin ligase, that acts upstream of *Bmp4* (Granata and Quaderi 2003). *Mid1* is initially expressed on both sides of the node but later is confined to the right side by Shh-dependent repression, where it induces *Bmp4* that in turn represses *Shh*. *Midline2* (*Mid2*), a close homolog of *Mid1*, shows a pattern of expression in the node similar to *Mid1* and can substitute for it during chick LR development (Granata et al. 2005). The redundancy between Mid1 and Mid2 may explain the absence of laterality defects in the human X-linked Opitz G/BBS (OMIM 300000) syndrome caused by mutations in *MID1* (Quaderi et al. 1997).

Pcl2 (*Polycomblike 2*) encodes a transcriptional repressor that can specifically repress Shh promoter activity in vitro and *Shh* expression in vivo (Wang et al. 2004). Coincident with the start of *Shh* asymmetric expression in the node, *Pcl2* expression is detected on the right side of the node and could therefore mediate Shh repression in this side of the node. Misexpression experiments indicate that Pcl2 is downstream of ActivinβB and Bmp4 (Wang et al. 2004).

In the chick, *ActivinβB*, *Mid*, *Bmp4*, and *Pcl2* arrange in an orderly fashion in a right-sided pathway aimed at abolishing *Shh* expression from this side of the node. This mechanism may reflect the importance of restricting *Shh* expression to the left side of the node where it induces a small domain of *Nodal* expression in the anterior left side of the node (perinodal domain; Figs. 7 and 8). In the avian embryo *Shh* is a crucial component of the left side pathway as demonstrated by gain- and loss-of-function experiments (Figs. 7 and 8; Levin et al. 1995; Pagan-Westphal and Tabin 1998).

Nodal is a member of TGFβ superfamily and is considered the left-side determinant since it is expressed in the left side of the embryo in all vertebrate species analyzed so far. *Nodal* expression in the small perinodal domain is not totally dependent on Shh induction because it is also independently activated by the Notch and Wnt/β-catenin signaling pathways.

Several recent publications have underscored the importance of the Notch signaling pathway in activating *Nodal* both in chick and mouse embryos (Krebs et al. 2003; Przemeck et al. 2003; Raya et al. 2004; Raya et al. 2003). In the chick embryo, before the appearance of the left side perinodal expression domain of *Nodal*, there is a boost of Notch activity on the left side of the node. The increase in Notch activity depends on a transient local accumulation of extracellular Ca^{2+} on the left side of Hensen's node from stage 4 to 6HH thought to be generated by the activity of the H^+/K^+ATPase pump. Two components of the Notch pathway, the ligand *Delta like 1* (*Dll1*), and the modulator *Lunatic fringe* (*Lfng*) show enhanced expression on the left side of the node at stage 5HH in a region overlapping the asymmetric perinodal expression of Nodal at stage 6HH (Fig. 9). Based on all these data, Raya et al. (2004) proposed a model, in which the activity of the H^+/K^+ATPase pump in the early embryo leads to a transitory rise in extracellular calcium on the left side of Hensen's node that strengthens Notch activity locally resulting in the activation of *Nodal* in this side of the node (Fig. 7). It is important to mention that Notch activity in the node is independent of Shh signaling (Raya et al. 2004; Raya et al. 2003). Mutations in some components of the Notch pathway in the mouse also give phenotype of defective laterality (Table 6). Particularly interesting is the mutation of *Delta like 1* (*Dll1*) (Krebs et al. 2003; Raya et al. 2003) which lacks *Nodal* expression in the node and, consequently in the LPM, indicating that Notch activity is required for the activation of Nodal also in the mouse (Krebs et al. 2003).

The possible relationship between the increase in intracellular calcium seen on the left side in the mouse node (McGrath and Brueckner 2003) and the rise in extracellular calcium described by Raya et al. (2004) is presently unexplored.

Wnt8c, a member of the Wnt family of growth factors that operates through β-catenin, is expressed on the right side of Hensen's node in the chick but despite its right side expression it functions as a left-side determinant (Fig. 9C; Levin 1998a; Rodriguez-Esteban et al. 2001). Misexpression experiments showed that Wnt8c, signaling through β-catenin, activates *Nodal* expression in a pathway independent of Shh. Since several Wnt ligands are symmetrically expressed in the node, it is likely that the sum of Wnt activities triggers Nodal but that this activity is repressed on the right side of the node. Indeed the adhesion molecule N-cadherin serves this function (Rodriguez-Esteban et al. 2001). N-cadherin is normally expressed on the right side of Hensen's node where it is thought to block *Nodal* expression by inhibiting the Wnt pathway in this side of the node (Fig. 7; Garcia-Castro et al. 2000; Rodriguez-Esteban et al. 2001). In the mouse, the study of the *Wnt3a* mutant has revealed that *Wnt3a* emanating from the posterior primitive streak functions at a distance to activate target genes in the node. One such target is *Nodal*, as indicated by its reduced expression in the posterior part of the node in the absence of *Wnt3a*, supporting therefore a role for Wnt signaling in Nodal induction similar in chick and mouse (Nakaya et al. 2005).

Left perinodal expression of *Nodal* in the chick is mediated by another mechanism dependent on the protein kinase A (PKA). Unexpectedly, Garcia-Castro et al. (2000) and Rodriguez-Esteban et al. (2001) found that the PKA positively regulated *Nodal* expression and that their function on the right side of the node was abolished by the protein kinase A inhibitor (PKI), which is expressed exclusively on the right side of Hensen's node (Kawakami and Nakanishi 2001; Rodriguez-Esteban et al. 2001).

Therefore, the restricted left-side perinodal expression of *Nodal* is under the control of multiple independent regulatory mechanisms that at least include the Shh, Notch, and Wnt pathways (Fig. 7).

In parallel with the left-sided pathways directed to inducing *Nodal*, the right-sided pathway previously mentioned simultaneously induces *Fgf8* on the right side of Hensen's node (Figs. 7 and 8; Boettger et al. 1999; Monsoro-Burq and Le Douarin 2001). *Fgf8*-signaling from the right side of the node prevents activation of *Nodal* on this side and transfers the information to the right LPM where it upregulates *Snail1* (Barrallo-Gimeno and Nieto 2005)—*Snai1* in Mouse Genome Informatics expression (Isaac et al. 1997).

In the mouse, asymmetric gene expression in the node appears at present to be less complex. A number of genes related to LR patterning are expressed in the mouse node including *Shh*, *Ihh* (*Indian Hedgehog*), *Nodal*, *Cfc1*, *Fgf8*, *Bmp7*, *Dand5*, *Lplunc1*, and *Gdf1*. Of these only *Nodal*, *Dand5*, and *Lplunc1* have been reported to show asymmetric expression (Collignon et al. 1996; Hou et al. 2004; Marques et al. 2004; Pearce et al. 1999). In stark contrast to that in the chick embryo, the mouse node does not show any asymmetry in *Shh* expression.

Initially, *Nodal* is symmetrically expressed in both sides of the mouse node around E7.5, but a few hours later (at E7.75) its expressions turns asymmetric as the left-sided domain becomes wider and stronger than the right side (Collignon et al. 1996; Lowe et al. 1996). From the 2–3-somite stage, *Nodal* is also expressed in the left LPM.

At the transcriptional level, node-specific *Nodal* expression is governed by a cis-acting regulatory element located 5′ upstream of the *Nodal* promoter. The node specific enhancer contains several scattered LEF1/TCF sequences, which mediate transactivation by β-catenin (Adachi et al. 1999; Norris and Robertson 1999; Saijoh et al. 1999) as well as Notch-responsive elements (Krebs et al. 2003; Raya et al. 2003), strongly supporting a role for Wnt and Notch signaling in *Nodal* regulation of expression, as indicated by the experiments in chick. However, the node enhancer alone drives symmetric *Nodal* expression in both sides of the node (Adachi et al. 1999; Norris and Robertson 1999; Saijoh et al. 1999).

Interestingly, recent studies have underscored the importance of cilia as organelles for signaling tansduction of Hedgehog genes (Huangfu and Anderson 2005; Liu et al. 2005; May et al. 2005; see Sect. 3.1.1). This raises the possibility that the role nodal cilia play in initiating asymmetric *Nodal* expression in the node may also be mediated by signaling by Hedgehog genes. However, Gli-binding sites have not been reported in the Nodal enhancers, making unlikely that Shh directly induces *Nodal* expression.

Lplunc1, a member of the BPI/PLUNC gene superfamily, is expressed in the crown cells of the node in a pattern very similar to that of *Nodal* (Hou et al. 2004). The asymmetric left-side expression of *Lplunc1* is maintained albeit at reduced levels in the absence of Nodal suggesting that it does not require Nodal signaling. Owing to the similarities between *Lplunc1* and other members of the family, it has been suggested that *Lplunc1* might bind and transfer lipids, an activity that is important for signaling molecules such as Shh and Wnt (Hou et al. 2004).

Dand5 (also *Cerl-2*) is a novel member of the DAN family of proteins, which also include Cerberus in the mouse and Caronte in the chick. These proteins have a cysteine knot domain that binds and blocks Nodal. Very interestingly, *Dand5* is asymmetrically expressed in the mouse node with expression stronger on the right, in a complementary pattern to that of Nodal (Marques et al. 2004; Pearce et al. 1999). *Dand5* knockout mice display multiple laterality defects including randomization of the LR axis with predominant bilateral *Nodal* expression. These defects have been interpreted as a result of the gene working to restrict the Nodal signaling pathway to the left and preventing its activity on the right side (Marques et al. 2004).

In summary, the present data indicate that in the chick the onset of asymmetry occurs in the peripheral tissues of the early gastrula embryo and is then conveyed to the node where a complicated network of gene expressions is established. In a next step the asymmetry is transferred from the node to the LPM. It should, therefore, be emphasized that the laterality information does not flow directly from the peripheral tissues to the mesoderm, but is transmitted through the node. In contrast, in the mouse the onset of asymmetry appears to be generated intrinsically within the node, rather than being imposed on it. Although asymmetries of gene expression are less prominent in the node in mouse than in chick, the final result is similar in the asymmetric, left-side biased, expression of Nodal in its perinodal domain.

2.3 Asymmetric Gene Expression in the Lateral Plate Mesoderm

From the node the asymmetric information has to be transferred to the LPM, the tissue that will actually contribute to the asymmetric development of organs (Fig. 10).

In the LPM the crucial expression is that of Nodal, specifically restricted to the left side (Collignon et al. 1996; Lowe et al. 1996). Nodal is considered the left-side determinant because modifications of its pattern of expression always correlate with *situs* alterations. Nodal expression in the left LPM is conserved in every species analyzed, from ascidians to mammals (Palmer 2004). Recently it has been reported that LR asymmetry in the sea urchin is regulated by Nodal signaling on the right side (Duboc et al. 2005). This last finding has been correlated with the evolutionary origin and transformation of the embryonic body axes.

Curiously, the expression of *Nodal* in the LPM, a decisive event in LR patterning, is highly dynamic and transient (Fig. 11). In the chick it starts at stage 7HH, lateral to the first formed somite, rapidly spreads anteriorly and posteriorly to cover the whole LPM (Fig. 11) and subsequently fades out so that by stage 10–11HH, *Nodal* expression only remains in the posterior part of the LPM. In mouse *Nodal* expression in the LPM is also very transient, lasting only for the period between the 2 and 6-somite stage (Collignon et al. 1996; Lowe et al. 1996).

A key question is how this asymmetric left-sided expression is controlled. Functional studies in the chick have demonstrated that Shh positively regulates *Nodal* expression, indicating that Nodal activation is Shh-dependent (Levin et al. 1995). Exogenous application of a source of SHH on the right side of the node leads to ectopic activation of *Nodal* on this side LPM. In contrast, abolition of Shh signaling by application of a monoclonal antibody inhibits *Nodal* expression (Levin et al. 1995). Both situations, bilateral or absent *Nodal* expression, result in randomization of heart looping (Fig. 12).

Given the distance between the node and the LPM, it was suggested that this induction could not be direct but mediated by a presumptive intermediate molecule (Pagan-Westphal and Tabin 1998). However, current evidence is compatible with Shh signaling controlling *Nodal* expression in the node, rather than directly in the LPM. From the node Nodal itself spreads to the LPM where it induces its own expression (see below).

In the mouse the transient asymmetric expression of *Nodal* in the left LPM requires the function of an intronic enhancer called <u>as</u>ymmetric <u>e</u>nhancer (ASE; Adachi et al. 1999; Norris and Robertson 1999; Saijoh et al. 1999). The ASE enhancer contains Foxh1 (also FAST1/2) and Smad2 (an intracellular mediator of Nodal signaling, see Sect. 3.2) binding sites that are crucial for the enhancer activity (Adachi et al. 1999; Norris and Robertson 1999; Saijoh et al. 1999). Therefore, Nodal signaling activates the ASE and induces its own expression. A second asymmetric enhancer called <u>l</u>eft-<u>s</u>ide-specific <u>e</u>nhancer (LSE) has recently been identified and, like the ASE enhancer also responds to Nodal (Saijoh et al. 2005).

Therefore, collectively the available data indicate that *Nodal* expression in the LPM is initiated by Nodal itself acting through Foxh1/Smad2–3 on the Nodal-responsive enhancer (Hamada et al. 2002; Yamamoto et al. 2003). The Nodal protein has to come from the node, the only embryo site expressing Nodal at the time, making *Nodal* expression in the node a prerequisite for subsequent LPM expression. Accordingly, loss of *Nodal* in the node results in absence of *Nodal* expression in the LPM (Table 6; Brennan et al. 2002; Saijoh et al. 2003) and ectopic induction of Nodal in the right LPM induces *Nodal* expression (Yamamoto et al. 2003). The mechanisms of Nodal propagation from the node to the LPM are not completely understood but may involve other members of the TGFβ superfamily such as Lefty (Meno et al. 2001).

The best-identified target of Nodal is the transcription factor Pitx2, involved in the control of organ laterality (Fig. 11; Amendt et al. 2000; Campione et al. 1999; Logan et al. 1998; Nielsen et al. 2001; Piedra et al. 1998; Rodriguez-Esteban et al. 1999; Ryan et al. 1998; Schneider et al. 1999; Yoshioka et al. 1998). *Nodal* expression, which is transient, imposes left-side identity on LPM cells by inducing *Pitx2* expression.

Pitx2 is a bicoid-related homeodomain transcription factor that is induced by Nodal in a pattern very similar to itself. Notably, *Pitx2* expression continues during handed morphogenesis and could therefore mediate the translation of molecular asymmetries to the organ level. *Pitx2* is expressed not only in the left LPM and derivatives, but also in other areas such as the cephalic mesoderm, the eye, and at later stages, in muscles (Logan et al. 1998; Piedra et al. 1998).

Pitx2 has three isoforms, generated by alternative splicing, called *Pitx2a*, *Pitx2b*, and *Pitx2c* (Schweickert et al. 2000). A fourth isoform has also been described in man (Cox et al. 2002). The *Pitx2* isoforms have differential patterns of expression, but the *Pitx2c* isoform is the only one expressed in the left LPM and derivatives.

Pitx2c expression in the left LPM is controlled by an ASE that contains Foxh1-Smad2/3 activation sites, like the *Nodal*ASE (Shiratori et al. 2001). The activation of *Nodal* expression in the LPM results in the onset of *Pitx2c* expression. Expression of *Pitx2c* in the LPM occurs in two steps (Shiratori et al. 2001). In a first step *Pitx2c* expression is initiated by Nodal, which activates the *Pitx2*ASE region. Once *Nodal* expression ceases, *Pitx2* expression is maintained by the transcription factor Nkx2.5, which is expressed in an organ-specific but left-side pattern (Shiratori et al. 2001). Accordingly, the way *Pitx2* expression is disturbed in laterality mutants replicates Nodal alteration of expression (Table 6). For instance, the *iv* mutation shows all patterns of expression of *Pitx2* in the LPM (Fig. 12) and the *inv* mutation shows right-side expression (Logan et al. 1998; Piedra et al. 1998). Null mutations that result in absence of *Nodal/Pitx2c* expression in the LPM include *Cfc1*, *Gdf1*, and *Cited2*, whose predominant phenotype is right isomerism. Null mutations that cause a predominant bilateral expression of *Nodal/Pitx2c*, such as *Lefty1/2*, give rise to a phenotype of left isomerism.

It is possible that Foxh1–Smad complexes and Nkx may recruit other coactivators. One possible candidate is *Cited2*, a member of the recently identified gene family termed CITED (CBP/p300 interacting transactivators with ED-rich termini; Dunwoodie et al. 1998) composed of transcriptional co-activators. *Cited2* is necessary for heart, adrenal, and nervous system development (Bamforth et al. 2001). Recently genetic and biochemical studies have shown that *Cited2* has a role in establishing left–right patterning, explaining the wide spectrum of cardiovascular malformations observed in the *Cited2* mutant (Bamforth et al. 2004; Weninger et al.

2005). *Cited2* null mice lack expression of the Nodal target genes *Nodal*, *Pitx2c*, and, *Lefty2* in the LPM and exhibit abnormal cardiac looping and right isomerism among other laterality defects (Bamforth et al. 2004; Weninger et al. 2005). Cited2 physically interacts and coactivates TFAP2 (transcription factor AP2) and endogenous Cited2 and TFAP2 are detected at the *Pitx2c* promoter and are capable of activating *Pitx2c* expression (Bamforth et al. 2004; Braganca et al. 2003). Therefore it seems that Cited2 may act as coactivator of Nodal-activated gene transcription in the LPM (Bamforth et al. 2004).

Besides *Nodal* and *Pitx2*, two genes encoding TGFβ antagonist, *Lefty2* and *Caronte* and two genes encoding transcription factors, *Nkx3.2* and *Snail1*, also show LR asymmetric patterns of expression in the LPM (Fig. 10).

Lefty2 and its closely related gene *Lefty1*, are members of the TGFβ superfamily and are characterized by the lack of a cysteine residue necessary for dimerization (see Sect. 3.2; Meno et al. 1996). Like *Nodal*, *Lefty2* is transiently expressed exclusively in the left LPM. At the 2-somite stage expression starts close to the node and subsequently spreads to cover the whole LPM up to the 5–6-somite stage when it becomes undetectable. Interestingly, *Lefty2* has an ASE similar to Nodal that is necessary and sufficient for its left side expression (Saijoh et al. 2000). The ASE is activated by Nodal, which also activates its own expression. Genetic evidence indicates that Lefty proteins act as antagonists of Nodal, thus leading to a model similar to the reaction-diffusion model, in which Nodal will induce its own expression and that of its antagonist (see Sect. 3.2; Hamada et al. 2002).

In the chick, Caronte (Car), a member of the Cerberus/Dan family of Bmp and Nodal antagonists, is expressed in the left LPM in a pattern similar to that of *Nodal* and *Lefty2* in the mouse (Fig. 13A; Rodriguez-Esteban et al. 1999; Yokouchi et al. 1999; Zhu et al. 1999). Caronte is positively regulated by Shh and is capable of inducing Nodal. Initially, Caronte was thought to be required for *Nodal* expression acting as an antagonist of Bmps (Rodriguez-Esteban et al. 1999; Yokouchi et al. 1999; Zhu et al. 1999); however, it was later shown that Bmp signaling in fact facilitated *Nodal* expression (Piedra and Ros 2002; Schlange et al. 2002). The function Car plays in the left LPM of the chick embryo is presently unknown and will require loss of function experiments for its determination.

Snail1, a zinc finger-type transcription factor, is asymmetrically expressed on the right LPM of chick embryos and is also involved in the control of *Pitx2* expression (Fig. 11D; Isaac et al. 1997; Patel et al. 1999). In the chick, loss-of-function of *Snail1* by antisense disruption experiments, indicates that *Snail1* may act by repressing *Pitx2* and that during normal development Nodal signaling represses *Snail1* activation on the left as another way of ensuring correct activation of *Pitx2* (Patel et al. 1999). In the mouse, *Snail1* homolog also shows asymmetric right-side expression in the LPM (Sefton et al. 1998).

Nkx3.2 (also Bapx1) is another transcription factor with asymmetric pattern of expression in the LPM. It belongs to the NK family of homeobox-containing genes (Tribioli and Lufkin 1997) and has been shown to direct spleen and pancreas development (Hecksher-Sorensen et al. 2004). Surprisingly, this factor was found to be expressed on opposite sides in chick and mouse LPM. In the chick, *Nkx3.2* is expressed in the left LPM and responds to left-side signals (Schneider et al. 1999). In contrast, in the mouse it is expressed in the right LPM and behaves as a right-side element, resulting accordingly modified in the *iv* and *inv* mutations (Rodriguez-Esteban et al. 1999). Presently there is no clear explanation for this divergence.

Another discrepancy between the chick and the mouse, not related to expression site but to function, occurs with Fgf8. In the chick, *Fgf8* is expressed on the right side of Hensen's node and works in the right-side pathway downstream of Bmp4 and upstream of Snail1 (Boettger et al. 1999; Isaac et al. 1997; Patel et al. 1999). However, in the mouse, expression of *Fgf8* is not asymmetric and it functions as a left-side determinant. This conclusion is based on the study of *Fgf8* hypomorphic mutant mice that exhibit right isomerism in half of the mutant embryos and do not express left-side determinants (Meyers and Martin 1999). Curiously in another mammal, the rabbit, Fgf8 functions as a right-side determinant, as in chick (Fischer et al. 2002). Since the rabbit blastocyst is very similar to that of the chick, it has been proposed that the spatial architecture of the blastocyst may account for the different function of *Fgf8* in different species (Fischer et al. 2002).

Despite the marked divergences in the early phases of LR specification, the expression of *Nodal* and its downstream gene *Pitx2* in the LPM is conserved in all vertebrates examined, constituting what has been called the "left–right phylotypic stage" (Yost 1998; Yost 2001). Expression of the *Lefty2* genes is also highly conserved in vertebrates, but so far it has not been identified in the chick. Of all the genes with asymmetric expression in the LPM, the expression of *Pitx2* fits temporally and spatially with functioning in the later stages of asymmetric organ morphogenesis.

Particularly interesting is the role *Pitx2* plays in heart development. During heart development, *Pitx2* is only expressed in the left side of the cardiac tube (Fig. 12A, B; Campione et al. 2001; Piedra et al. 1998). During the process of bending and rotating, *Pitx2c* expression remains in the same left-sided cells that acquire a ventral position at ventricular level, as demonstrated by fate-mapping studies (Campione et al. 2001). Expression of *Pitx2* extends from the left side of the inflow tract to the outflow tract and into the secondary heart field. With further development, *Pitx2c* expression becomes confined to the left part of the inflow tract and atrioventricular canal. Finally expression of *Pitx2c* declines during fetal stages to become undetectable by E16.5 onward. Interestingly, *Pitx2c* is expressed in the presumptive heart field in a subpopulation of left branchial arch and splanchnic mesoderm (Liu et al. 2002).

Overexpression of *Pitx2* in the right LPM of chick and *Xenopus* showed randomization of cardiac looping (Logan et al. 1998; Ryan et al. 1998). However, cardiac looping is normal both in the *Pitx2* null mutant and in the *Pitx2c* isoform-specific mutant, although the heart exhibits multiple defects usually associated with heterotaxia (Gage et al. 1999; Kitamura et al. 1999; Liu et al. 2001; Liu et al. 2002; Lu et al. 1999). However the heart phenotypes are not the same in the two types of *Pitx2* mutants. Despite normal looping, the heart in the absence of any *Pitx2* isoform exhibits severe malformations that have dissimilarities with regard to those occurring specifically in the absence of *Pitx2c* (Gage et al. 1999; Kitamura et al. 1999; Liu et al. 2001; Liu et al. 2002; Lu et al. 1999). The isoform-specific deletion of *Pitx2c* reveals that it functions to regulate asymmetric branchial arch remodeling and to pattern the outflow tract (Liu et al. 2002).

In humans, Axenfeld-Rieger syndrome (OMIM 180500) is caused by haploinsufficiency of the *PITX2* gene and is characterized by dental hypoplasia and ocular, craniofacial, and umbilical anomalies (Semina et al. 1996). With lower frequency, cardiac and limb malformations have been reported to form part of the syndrome (Amendt et al. 2000). Surprisingly, this syndrome does not carry laterality defects, probably due to the fact that a reduced dose of Pitx2 (one allele) may suffice to fulfill this function (Liu et al. 2001). In this regard it is interesting to mention that different organs have distinct requirements for Pitx2c dosage for correct development (Liu et al. 2001). For instance, the cardiac atria require low *Pitx2*c levels whereas the duodenum and lungs use higher doses for normal development.

2.4 Role of the Midline in Establishing Left–Right Asymmetry

Morphologically, the early embryo is symmetric along the AP axis dividing the embryo into two halves, left and right. This axis is considered the midline of the embryo and is occupied by the primitive streak in the chick and mouse and by the dorsal midline in fish and frogs. After gastrulation, the midline is morphologically formed by the notochord, which is a derivative of the node, and the neural tube. The midline tissue forms concomitantly with the establishment of LR patterning. It is of considerable interest that during gastrulation very few cells cross the midline from one side of the embryo to the other as they migrate through the primitive streak (Levy and Khaner 1998).

The first evidence of the importance of the midline in establishing LR asymmetries came from experiments in *Xenopus* aimed at preventing the development of anterior dorsal structures (Danos and Yost 1995; Danos and Yost 1996). These embryos, which lacked both the head and the notochord, also showed abnormal organ situs that correlated with the extent of the notochord defect.

Further evidence came from the finding that zebrafish and *Xenopus* mutants with defective midline structures also had associated laterality defects. In zebrafish mutants *no tail* (*ntl*) and *floating head* (*flh*; Rebagliati et al. 1998; Sampath et al. 1998) are characterized by the absence of notochord and these mutant embryos exhibit randomization of heart situs. These two mutants express *Pitx2* bilaterally in the left and right LPM. The *ntl* harbors a mutation at the *T* (*Brachyury*) locus (Danos and Yost 1996). T (Brachyury) is a transcription factor that is essential for mesoderm formation and is normally expressed in the notochord. *T* (*Brachyury*) mutations in the mouse also affect LR development because of interference with midline maintenance (King et al. 1998). In zebrafish several mutations with laterality phenotypes have midline defects (Chen et al. 1997).

Besides the *T* gene mutation, several mutations in mice showing laterality defects have been explained by the concurrent existence of midline defects (Table 6). One such mutation is *no turning* (*nt*), characterized by failure of embryonic turning, defective caudal development, notochord degeneration, and randomization of heart *situs* (Melloy et al. 1998). This mutation, autosomal recessive and lethal before E11.5, perturbs the asymmetric expression of *Nodal*, which is predominantly bilateral in the LPM although all other patterns of expression, namely absent, right, and left, may appear. The LR phenotypic defects are explained by midline alterations.

Foxa2 (Hepatocyte nuclear factor 3β, HNF3β) is part of a family of fork head transcription factors first identified as liver-specific (Lai et al. 1990). It is also expressed in early embryogenesis particularly in the node, notochord, and floorplate (Ang et al. 1993). Genetic disruption and misexpression experiments have suggested that it is required for node and notochord formation and thus involved in patterning the midline of the embryo (Ang and Rossant 1994; Sasaki and Hogan 1994; Weinstein et al. 1994). The analysis of chimeric embryos using tetraploid embryo-ES cell aggregations, which develop further than mutant embryos, have revealed its involvement in LR patterning (Dufort et al. 1998). Chimeric embryos lacked *Nodal* expression and showed absent or bilateral expression of *Lefty2*. Accordingly there is a failure in heart looping. Since mutant embryos do not form the node or the notochord, a simple explanation of the phenotype is the absence of midline structures.

Foxh1 (*FAST1/2*) is a gene closely related to *Foxa2* necessary for the formation of the node, notochord, and prechordal plate mesoderm (Hoodless et al. 2001; Yamamoto et al. 2001). Indeed the phenotype of *Foxh1* null mice is similar to that of *Foxa2* mutant lacking expression of left-side markers, and it has been demonstrated that *Foxh1* is required for *Foxa2* expression (Hoodless et al. 2001). *Foxh1* is a major transducer of Nodal signaling (Sect. 3.2).

Shh mutant mice show multiple defects in the midline including the rapid degeneration of the notochord and the absence of floor plate (Chiang et al. 1996), corresponding to the *Shh* pattern of expression and function in ventral neural tube development (Echelard et al. 1993; Ericson et al. 1995). In high contrast with Shh functioning as a left-side determinant in chick, *Shh* mutants show pulmonary left-isomerism due to bilateral expression of left-sided markers (Izraeli et al. 1999; Litingtung et al. 1998; Meyers and Martin 1999; Motoyama et al. 1998; Tsukui et al. 1999; Zhang et al. 2001). The phenotype of the *Shh* mutant can be explained by the midline deficiencies, but other genetic studies support a more direct role for Shh signaling in the generation of LR asymmetries (see Sect. 3.1).

The *Stil* (*Scl/Tal1 interrupting locus,* also *SIL*) gene is an early-response gene with ubiquitous expression in proliferating cells (Aplan et al. 1990; Izraeli et al. 1997). In the absence of *Stil* (Izraeli et al. 2001; Izraeli et al. 1999), embryos display prominent midline neural tube defects and randomization of heart *situs*. The molecular characterization of this mutant revealed that *Shh* signaling in the midline was blocked, with greatly reduced *Ptc* (*Patched*) and *Gli1* expression. However, a midline block in *Shh* signaling cannot completely explain the *Stil* phenotype since

Shh mutants have normal cardiac *situs*. A careful analysis of the pattern of expression of *Nodal*, *Lefty2*, and *Pitx2* in *Shh* and *Stil* mutants led Izraeli and colleagues to propose that the phenotype differences could be explained by temporal and spatial differences in the alterations of the pattern of expression of these left-side markers (Izraeli et al. 1999). There is genetic evidence that Stil is required for the Shh transduction pathway downstream of Ptc, although its precise biochemical role remains to be determined (Izraeli et al. 2001).

In the mouse, there are several mutations that course with associated midline and laterality defects (Table 6). Genetic disruption of the *Mgat1* gene, encoding an enzyme involved in the biosynthesis of complex oligosaccharides, also results in randomization of heart situs (Metzler et al. 1994). Curiously this mutation also causes dorsal midline defects that have been considered to explain the phenotype. *Rotatin* (*Rttn*) is also a novel gene involved in LR specification through its function in notochord development (Faisst et al. 2002).

This increasing number of mutants, together with experiments of ablation of midline structures, indicates that abnormal development of the notochord or degenerative loss of the notochord lie at the root of laterality defects, which are mainly mediated by bilateral expression of *Nodal* and *Pitx2*, instead of the normal left-side expression. All this evidence suggests that a midline structure is essential for the generation of correct LR asymmetries, acting as a barrier that prevents signaling on one side of the embryo from affecting the other side. The midline would be responsible for maintaining the independence of the left- and right-side information pathways. In this regard, it should be noted that there is no mixing of the left and right lateral plate mesodermal populations, since during gastrulation the number of cells crossing the midline is minimal (Levy and Khaner 1998).

Curiously, some of the mutants with a defective midline show bilateral expression of left-sided genes while others show absence of expression. A possible explanation for this difference may lie in a different degree of disruption of the node. When the disruption of the node is early and severe no induction of left-sided genes occurs and cannot, therefore, pass to the other side. For example, *T* mutants may fail to initiate the cascade of gene expression required for the establishment of left identity. In summary, the specific onset and extended duration of *Nodal*, *Lefty2*, and *Pitx2* expressions in the whole of, or in part of, the right LPM provide the key to explaining the variable LR phenotype of mutants with midline defects (Izraeli et al. 1999).

Although the concept of the midline as a barrier is the prevailing model, alternative models have been proposed. For instance, the observation that midline structures repressed a Nodal-related gene on the right LPM of *Xenopus*, led to the hypothesis that the midline could act as a repressor of the signaling pathway on the right side of the embryo (Lohr et al. 1997). This view is also supported by *Stil* mutants that show an early onset of *Lefty2* expression on the right LPM, before normal left-sided expression (Izraeli et al. 1999).

The study of the laterality defects in conjoined twins fits nicely with the idea that a barrier is absolutely essential to confine signaling pathways to the appropriate site. The study of spontaneously occurring conjoined twins in chicks (Levin et al. 1996), together with the available knowledge in humans, indicates that the signaling cascades from one embryo could influence the other embryo, depending on the geometry of the duplicated primitive streaks. The midline barrier model provides a good framework for interpreting studies of twins (see Sect. 2.5).

The distinction as to whether the midline acts as a physical barrier (morphological entity) or as some kind of molecular barrier came from the study of the genetic disruption of the *Lefty1* gene (*EBAF* in humans; Meno et al. 1998). *Lefty1* is expressed in the left side of the presumptive floor plate and/or the notochord in all vertebrates examined (Fig. 13B; Meno et al. 1998; Piedra and Ros 2002; Rodriguez-Esteban et al. 1999). Its expression is transient, lasting from the 2-somite to the 5–6-somite stage (Meno et al. 1998). *Lefty1* expression starts shortly after Nodal activation of expression on the LPM and genetic evidence indicates that it is regulated by Nodal from the LPM (Hamada et al. 2002; Yamamoto et al. 2003). In the absence of *Lefty1* the notochord and neural tube form normally, but left-sided expressed genes such as *Nodal* and *Lefty2* propagate across the midline and are found bilaterally. Since *Lefty1* is a divergent member of the superfamily of TGFβ that binds Nodal, Meno and coworkers proposed that the role of Lefty1 protein is to serve as a molecular barrier that prevents diffusion of a molecule regulating expression of *Lefty2* and *Nodal* (Meno et al. 1998). It was later discovered that this molecule was Nodal itself (Saijoh et al. 2003; Yamamoto et al. 2003).

It is worth mentioning that two mutations that affect components of the Bmp signaling pathway, the *Smad5* and *Acvr1* mutations (see Sect. 3.3), show absence of *Lefty1* expression in an otherwise normal midline. In agreement with the proposed function for Lefty1, these two mutants express left-side markers bilaterally (Chang et al. 2000; Kishigami et al. 2004). This finding indicates that Bmp signaling, like that of Nodal, may positively regulate *Lefty1* in the mouse midline, contrary to what has been shown to occur in the chick (Piedra and Ros 2002; Yokouchi et al. 1999).

Recently (Kelly et al. 2002) have defined a functional midline in chick formed by the central dorsal cells in the primitive streak that express *Fgf8* and *Brachyury*. These cells are fated to die yet maintain their position in the midline (see Fig. 6A,B). Most interestingly, loss of midline cell death is followed by randomized heart looping, which indicates a role in LR asymmetry regulation, possibly by acting as a physical barrier since the dead cells remain in the streak. It remains to be shown whether this cell death is also present in the primitive streak in species other than the chick.

The importance of the midline in establishing normal LR patterning is also known in human pathology. Of particular interest are some retrospective clinical studies that revealed a significant association between laterality defects and anomalies of the spine and other midline structures (Goldstein et al. 1998; Martinez-Frias et al. 1995; Ticho et al. 2000). Laterality defects were associated with congenital midline anomalies in 31% of patients studied by Goldstein et al. (1998), and this figure rose to 38% in a postmortem study of cases based on complete autopsy reports (Ticho et al. 2000).

2.5 Laterality Defects in Conjoined Twins

It was an ancient observation that some monozygotic and conjoined human twins exhibited alterations of laterality (Aird 1959; Torgersen 1950). Alterations of laterality were also reported as a consistent finding in experiments yielding conjoined embryos, such as those produced by medial constriction in early new embryos or by ectopic induction of a new organizer in *Xenopus* (Nascone and Mercola 1997; Spemann and Falkenberg 1919). While the twin on the left was completely normal, the twin on the right exhibited randomization of situs. In human conjoined twins it was also known that the right twin was the one that most frequently showed defects in laterality. Conjoined twins are popularly known as "Siamese twins" because of the two well-known brothers, born in (then) Siam in 1811; they were joined at the lower sternum (*xiphopagus*), lived for 63 years and even had children.

Conjoined twins reflect a twinning event occurring at the time of gastrulation (13–14 days postfertilization in humans) and leading to an incomplete split of the anterior–posterior axis so that the pair of twins remain joined by the part of the axis that fails to split. Therefore, conjoined twins have to be monoamniotic monochorionic twins. The joining tissue may only be superficial subcutaneous tissue, but may include cartilage and bone or even the sharing of vital viscera.

Conjoined twins can be classified into symmetric and asymmetric types accordingly to the relative proportion between the two twins. They also receive different names depending on the anatomical region of fusion (Table 3). In humans the most common region is the thorax, *thoracopagus* (Kaufman 2004).

In their excellent study, Levin and colleagues provided a reasonable explanation for the laterality defects seen in conjoined twins as well as correlating the type of defect with the different types of twins (Levin et al. 1996). Their explanation is based on the possible interaction/interference of LR signaling in one twin with that of the adjacent twin. The way this interference occurs depends on the spatial arrangement

Table 3 Nomenclature and site of fusion of conjoined twins

Name	Site of fusion
Prosopagus	Face
Cephalopagus	Head
Thoracopagus	Thorax
Sternopagus	Sternon
Xiphopagus	Apophysis xiphoids
Omphalopagus	Umbilical region
Ileopagus	Parietal bone
Pygopagus	Sacrococcygeus
Ischiopagus	Ischiatic region

of the two primitive streaks in the blastocyst (Fig. 14). The twinning event may result in the formation of two streaks opposed at their anterior or posterior end (*craniopagus* or *isquiopagus*). In both cases, the signaling cascades arising around the nodes do not interfere with each other and therefore the twins present *situs solitus*. This is confirmed by the lack of laterality defects in human *craniopagus* and *isquiopagus* and by the study of multiple spontaneous conjoined chick embryos (Levin et al. 1996). These findings clearly show that in a single blastocyst the specification of left and right sides can occur on either side confirming the notion that the specification of the LR axis takes place after determination of the AP and DV axes.

When the twinning event results in the formation of two parallel streaks, the right-sided inhibitors of the left-side signaling cascade in the left twin may also inhibit the left-side cascade in the right twin, if they are close enough. The right embryo, therefore, develops randomization of *situs*. This explanation is confirmed by observations in spontaneously occurring conjoined chick and human twins and by the analysis of *Nodal* expression patterns in chick conjoined twins (Levin et al. 1996). In cases of oblique convergent streaks, the left-side signaling cascade in the right twin may induce left-sided genes on the right side of the adjacent twin and in these cases it is the left twin that shows LR alterations (Fig. 14).

2.6 Retinoic Acid and Left–Right Patterning

Multiple studies have shown that retinoic acid (RA), the active metabolite of vitamin A, is an important factor during embryogenesis and particularly in LR development. During early embryonic development either excess or deficit of vitamin A results in alteration in the development of the LR axis. Administration of excess RA during gastrulation results in abnormal heart looping in the chick embryo (Dickman and Smith 1996; Smith et al. 1997) and heterotaxia in hamsters, mice, and rats (Sasaki et al. 1990; Shenefelt 1972; Wasiak and Lohnes 1999).

Several observations indicate that administration of excess RA can randomize the establishment of the LR axis during gastrulation. Deprival of vitamin A, thereby causing a deficit in RA, also affects LR patterning since it results in *situs inversus* that can be rescued by exogenous administration of Vitamin A (Dersch and Zile 1993; Twal et al. 1995).

Molecular analysis after treatment with excess RA during gastrulation shows bilateral expression of *Nodal* and *Pitx2* in mice, chick, *Xenopus*, and zebrafish (Chazaud et al. 1999; Rodriguez-Esteban et al. 2001; Tsukui et al. 1999; Wasiak and Lohnes 1999). Conversely, RA antagonism inhibits the expression of these genes. Therefore, it has been suggested that RA is an upstream factor controlling left-side determinants and that the nascent mesoderm constitutes a predominant target of RA, probably acting through *Lefty1*, a gene also identified as an RA-responsive gene (Chazaud et al. 1999).

Surprisingly, the recent study of *Aldh1a2*- (also *Raldh2*-) deficient mice (deficient in RA biosynthesis) shows that these mutants, although showing a phenotype of cardiac defects, exhibit normal expression of molecular markers of laterality (Niederreither et al. 2001). Lack of RA synthesis appears to severely affect atrial outgrowth as well as cardiomyocyte differentiation. Remarkably, *Aldh1a2*-deficient mice show a right-side delay in somite formation, from the 8- to 15-somite stages (Vermot et al. 2005). This asymmetry in somite formation is caused by the influence of the LR pathways (Kawakami et al. 2005; Vermot et al. 2005). Therefore, it has been suggested that RA acts as a buffering mechanism to dim the influence of the left side cascade and allow the harmonic development of somites in both sides of the body in this way permitting that patterning of the exterior body wall occurs symmetrically (Tabin 2005; Vermot and Pourquie 2005).

3 Signaling Pathways with a Predominant Role in Left–Right Patterning

We will consider in this section some of the signaling pathways with a proven critical role in establishing LR asymmetries.

3.1 Sonic Hedgehog Signaling and Left–Right Asymmetry

Shh is one of the three vertebrate *hedgehog* genes. The other two are *Desert hedgehog* (*Dhh*) and *Indian hedgehog* (*Ihh*; Echelard et al. 1993). *Shh* is a potent signaling molecule with multiple roles during embryonic development, including patterning in the central nervous system and in the limb. As already described, Shh is also involved in LR development, as a left-side determinant in the chick and in the mouse (Levin et al. 1995; Levin et al. 1997; Zhang et al. 2001) although it has also been considered a right-side determinant in mouse (Meyers and Martin 1999).

Both in chick and mouse embryos, *Shh* is expressed in the node. In the chick, *Shh* transcription in Hensen's node is observed bilaterally and symmetric at stage 4HH, but becomes restricted to the left side during stage 5HH (Fig. 8A). The left-sided asymmetric expression remains up to stage 7HH (Levin et al. 1995). In the mouse embryo, *Shh* is transcribed throughout the node with no appreciable asymmetry from the head fold to the 6-somite stage (Collignon et al. 1996). Remarkably, *Ihh* is also weakly expressed in the periphery of the mouse node, overlapping with *Shh* expression in its posterior part at E7.75–8 (Zhang et al. 2001). As with *Shh*, *Ihh* expression in the mouse node shows no LR asymmetry (Collignon et al. 1996; Zhang et al. 2001).

Besides the node, *Shh* is also expressed in midline tissues such as the notochord, floor plate and gut endoderm (Echelard et al. 1993), where it plays important roles in patterning.

Reception and transduction of signaling by Hedgehog family members is a complex process (Hooper and Scott 2005). The Shh receptor is a transmembrane protein called Ptc, which, in the absence of Shh, inactivates another transmembrane protein called Smoothened (Smo; Kalderon 2000). Signal transduction downstream of Smo is complex and not completely understood, but it is known that the main transcriptional mediators of Shh are the Gli genes (Ingham and McMahon 2001; Lee et al. 1997; Ruiz i Altaba 1998; Sasaki et al. 1997). Three Gli genes, *Gli1*, *Gli2*, and *Gli3*, have been identified in vertebrates. The Gli proteins are multifunctional transcription factors that may act as transcriptional activators or transcriptional repressors. Gli1 and Gli2 are primarily transcriptional activators of Shh target genes, while Gli3 is primarily a transcriptional repressor of Shh targets. Interestingly, the three Gli genes show specific and complementary patterns of expression in the chick Hensen's node, with *Gli1* and *Gli2* being expressed on the left and *Gli3* on the right side of the node (Granata and Quaderi 2005).

In humans, *Shh* haploinsufficiency results in holoprosencephaly (Belloni et al. 1996; Nanni et al. 1999; Roessler et al. 1997). This is a severe malformation caused by a defective division of the prosencephalon into the two cerebral hemispheres. There are different degrees of severity accompanied by corresponding severe skull and facial defects. The most severe cases are incompatible with life and closely replicate the phenotype seen in *Shh* null mice, characterized by a single-lobed brain, cyclopia, and proboscides (Chiang et al. 1996).

The *Shh* mutant presents multiple defects in the midline including failure to develop the floorplate and notochord degeneration (Chiang et al. 1996). Regarding LR symmetry, the *Shh* mutant also displays important defects such as left pulmonary isomerism with bilateral monolobed lungs that are severely hypoplastic due to impaired branching morphogenesis (Litingtung et al. 1998; Motoyama et al. 1998). However, cardiac looping is normal or shows only minor deficiencies (Izraeli et al. 1999; Meyers and Martin 1999; Tsukui et al. 1999; Zhang et al. 2001). The results of molecular analysis of patterns of *Nodal* and *Lefty2* expression in *Shh* null mice have varied depending on the particular studies, and have been reported as mostly left-sided or bilateral. Bilateral *Pitx2* expression in the LMP is a consistent finding (Izraeli et al. 1999; Meyers and Martin 1999; Tsukui et al. 1999; Zhang et al. 2001). These data may be interpreted as Shh functioning in the mouse to repress expression of left-side determinants on the right, thus acting as a right-side determinant (Meyers and Martin 1999).

Alternatively, the laterality defects seen in the *Shh* mutant may be mediated indirectly by the severe defects in midline structures and the lack of *Lefty1* expression (Izraeli et al. 1999; Zhang et al. 2001). In this respect, the study of total absence of Hedgehog genes caused by removal of *Smo* has shed light on this topic and suggests a direct role for Hedgehog signaling in LR patterning in the mouse node, similar to such events in the chick (Zhang et al. 2001).

In mammals there is only one *Smo* gene that plays an important role in transducing Hedgehog gene signaling (Akiyama et al. 1997). Mutations that constitutively activate *Smo* are responsible for some forms of basal cell carcinoma, because Shh target genes are constitutively expressed (Xie et al. 1998). Very interestingly, in *Shh;Ihh* double mutants, as well as in the *Smo* mutants, *Nodal* is expressed in the node, although with variable intensity and asymmetry, but is absent from the left LPM. Accordingly, the expression of left-sided markers is not turned on. This mutational analysis reveals that Shh and Ihh have redundant signaling functions in the establishment of LR asymmetries and that the mouse may use pathways similar to those in the chick in establishing the LR pathway.

Gdf1, a member of the Vg1/GDF subset of TGF-β superfamily ligands, is normally expressed in the node, but is absent in *Smo* and in *Shh;Ihh* compound mutants. Since *Gdf1* is required for the expression of *Nodal*, *Lefty2*, and *Pitx2* in the LPM (Rankin et al. 2000; Wall et al. 2000), a model has been proposed by Zhang et al. (2001) in which Shh and/or Ihh are required in the node for activation of *Gdf1*, which in turn activates *Nodal* expression in the LPM. In support of this view the phenotype of *Gdf1* mutants with regard to LR defects is very similar to that of *Nodal* hypomorphs.

Other modulators of Shh signaling, including *Dispatched* (*Disp1*), *Suppressor of fused* (*Sufu*), and the previously mentioned *Stil* show a LR phenotype when mutated (Table 6) (Cooper et al. 2005; Izraeli et al. 2001; Ma et al. 2002). This is another indication of the possible participation of Shh signaling in the specification of LR patterning.

3.1.1 Intraflagellar Transport Proteins and Shh Signaling

Interestingly, in an ethylnitrosourea-mutagenesis screen several genes were identified because of a phenotype similar to *Shh* loss-of-function (Huangfu et al. 2003). Two of these mutations disrupt genes homologous to Intraflagellar Transport (IFT) proteins, first characterized in *Chlamydomonas* for their role in the growth and maintenance of flagella (Rosenbaum and Witman 2002; see Appendix). These two genes are *Ift172* (also called *wimple*) and *Ift88* (also called *polaris* and *flexo*). The null phenotype of *Ift172* and *Ift88* are similar to each other and to that seen in defective Shh signaling, and include lack of ventral neural types. These two mutations also have laterality defects, showing randomization of heart looping as well as bilateral expression of left-sided markers (Huangfu et al. 2003; Murcia et al. 2000). The explanation for the LR phenotype is provided by the absence of cilia in the node. However, besides their role in ciliogenesis, IFT proteins have also been shown to play an integral role in the Shh pathway downstream of Ptc1 and Smo and upstream of Gli proteins, as indicated by genetic studies (Haycraft et al. 2005; Liu et al. 2005). In the absence of IFTs both activating and repressing functions of Gli proteins are hampered (Huangfu and Anderson 2005; Liu et al. 2005). Interestingly, loss of either the retrograde dynein motor *Dnahc2* (*dynein axonemal heavy chain 2*), or loss of the anterograde kinesin motor *Kif3a*, causes defective Shh signaling transduction and gives the same phenotype of Shh loss-of-function in the neural tube (Huangfu and Anderson 2005). All these data strongly indicate that cilia are organelles essential for transducing signaling of Hedgehog genes. In this regard it is interesting to note that several components of the signaling pathway of hedgehog genes have recently been reported to localize to cilia. They include Smo, Sufu, and the three Gli proteins (Corbit et al. 2005; Haycraft et al. 2005; May et al. 2005). Together all these studies suggest a complex relationship between cilia and transduction of signaling by hedgehog genes, raising the possibility that the protein machinery regulating hedgehog activity localizes to the cilia. They also raise the question of whether the LR phenotypes observed in absence of nodal cilia could also be mediated by defective Shh signaling.

3.2
Nodal Signaling and Left–Right Asymmetry

Nodal is a member of the superfamily of TGFβ cytokines, distantly related to the Bmp subfamily, and plays a central role in establishing LR patterning. In the early stages, Nodal genes have essential roles in mesoderm and endoderm induction (Schier and Shen 2000). Nodal genes include human; mouse and chick *Nodal*; zebrafish *Cyclops*, *Squint*, and *Southpaw*; and *Xenopus Xnr1*, *Xnr2*, *Xnr4*, *Xnr5*, and *Xnr6*. Nodal genes have also been identified in ascidians and in *Amphioxus* (*AmphiNodal*) but not in *Drosophila* or *C. Elegans* (Schier 2003).

Nodal is very transiently expressed in the left LPM of all vertebrates analyzed to date and represents the earliest asymmetric expression that is clearly conserved among vertebrates. A host of experiments indicate that normal left-sided expression of *Nodal* in the LPM is required for normal LR patterning. Mutations in the mouse that alter the pattern of *Nodal* expression result in concordant alterations of organ *situs*. Thus, the *iv* mutation exhibits all possible patterns of expression, namely bilateral, absent, left-sided, and right-sided, and is followed by randomization of *situs* and *heterotaxia* (Collignon et al. 1996; Lowe et al. 1996). The *inv* mutation results in *situs inversus* in homozygosis and these mice express right-sided Nodal (Collignon et al. 1996).

Like other members of the TGFβ family, Nodal signaling is achieved through binding to two cells surface serine–threonine receptor kinases, classified as type I and type II according to structural and functional criteria. In the absence of ligand, receptor types I and II remain as homodimers. Binding of the ligand causes receptor type I and type II to join together and form a hetero-tetrameric receptor complex (Shi and Massague 2003). The type II receptor phosphorylates the type I receptor to activate its kinase activity, which subsequently phosphorylates downstream targets such as the Smad proteins (Fig. 15; Table 4; Shi and Massague 2003).

The Smad proteins are a group of eight intracellular proteins, Smad1 to Smad8 (Massague and Chen 2000). They play a fundamental role in transducing signaling of some members of the TGFβ family including Nodal, Activin, and Bmp signaling. Because of their function, Smad proteins have been classified into three classes: receptor-regulated Smads (R-Smads), common-partner Smads (co-Smads), and inhibitory Smads (I-Smads). Two R-Smads (Smad2 and Smad3) have been shown to mediate Nodal signaling resulting phosphorylated upon activation of the receptor complex by ligand binding. Phosphorylation of the R-Smads causes a conformational change in the protein that allows them to associate with the co-Smad (Smad4), thus forming a complex that is translocated to the nucleus, where it

Table 4 Nodal signaling receptors and specific Smads

Nodal receptors			
Type I	Synonyms	Type II	Smads
Acvr1b	Alk4	Acvr2a Acvr2b	2, 3
Acvr1c	Alk7	Acvr2a Acvr2b	2, 3

activates or represses Nodal target genes (Shi and Massague 2003). The I-Smads (Smads 6 and 7) are able to bind to receptor type I in competition with R-Smad and, therefore, work against Nodal signaling. The transcription factor Foxh1 is also an intracellular effector of Nodal signaling (Yamamoto et al. 2001). Foxh1 forms complexes with phosphorylated Smad2 or Smad3 that activate Nodal target genes such as *Lefty2* (Saijoh et al. 2000).

Mutations in different elements involved in Nodal transduction of signaling confirm Nodal as a left-side determinant. *Nodal* mutants die very early due to Nodal requirement during early anteroposterior patterning and gastrulation (Zhou et al. 1993). However, several conditional *Nodal* mutants have been generated to analyze its role in LR development (Table 6). *Nodal* hypomorphs that develop normally up to the early somite stage show residual *Nodal* expression in the node but total absence in the left LPM (Saijoh et al. 2003). Accordingly, these mutants lack expression of left-side markers, as well as absence of *Lefty1* expression in the midline and develop a corresponding phenotype of right isomerism. Mutant mice specifically lacking Nodal in the left LPM or in the node also show a phenotype of right isomerism (Brennan et al. 2002; Norris et al. 2002) and indicate that Nodal from the node is necessary for *Nodal* expression in the LPM.

As mentioned, Foxh1 is an important transducer of Nodal signaling but *Foxh1* mutants die very early with a very defective midline (Hoodless et al. 2001; Yamamoto et al. 2001). Very interestingly, mice specifically lacking Foxh1 in the LPM, fail to express *Nodal*, *Lefty2*, and *Pitx2* and show right isomerism (Yamamoto et al. 2003). All the above mentioned studies indicate that in the absence of Nodal signaling in the LPM, the embryo develops right isomerism.

Smad2 mutant mice exhibit early patterning defects explained by a lack of Nodal signaling (Nomura and Li 1998; Waldrip et al. 1998), but compound mutations of *Nodal* and *Smad2* (double heterozygous) also result in dextro-isomerism (Nomura and Li 1998).

Nodal is secreted as a proprotein whose maturation involves the function of the proprotein convertases Furin and Pcsk6 (Beck et al. 2002). Surprisingly, *Furin* mutants, similarly to *Pcsk6* mutants, have bilateral activation of *Lefty2* and *Pitx2*. This result does not fit with the expected deficient Nodal signaling in *Furin* and *Pcsk6* mutants. Since these two proproteins also function in the maturation of

Table 5 BMP-signaling receptors and specific Smads

Type I	Synonyms	Type II	Ligand	Smads
Bmpr1a	Alk3	Bmpr2 Acvr2a	Bmp2, 4	1, 5, 8
Bmpr1b	Alk6	Bmpr2 Acvr2a Acvr2b	Bmp2, 4, 7 Gdf5	1, 5, 8
Acvr1	Alk2	Bmpr2 Acvr2a	Bmp2, 4, 7	1, 5

other members of the TGFβ family, it is difficult to precisely dissect the pathways affected in their absence (Constam and Robertson 2000a; Constam and Robertson 2000b; Roebroek et al. 1998). An added level of complexity described for Nodal signaling is the possibility of forming heterodimers with other members of the TGFβ superfamily, such as Bmp4 and Bmp7, which may modify its signaling activity (Yeo and Whitman 2001).

Besides randomization of the heart *situs*, mice mutant for the Nodal receptor *Acvr2b* (Table 4) exhibit right atrial and pulmonary isomerism, indicative of loss of Nodal signaling (Oh and Li 1997). But Acvr2b may also mediate Bmp signaling, particularly Bmp4 (Table 5; Chang et al. 1997), making it possible that several pathways are affected in this mutation. Compound mutations in *Acvr2b* and *Acvr2a* present laterality defects (Oh and Li 1997; Oh et al. 2002; Song et al. 1999).

Signaling by Nodal also requires the participation of EGF–CFC coreceptors that are small extracellular glycosylated proteins (Cheng et al. 2003; Gritsman et al. 1999; Schier and Shen 2000; Whitman and Mercola 2001; Yeo and Whitman 2001).

Members of the family of EGF–CFC include *Cripto* and *Cfc1* (*Cryptic*) in humans and mouse (Bamford et al. 2000; Ciccodicola et al. 1989; Dono et al. 1991; Dono et al. 1993; Shen et al. 1997), *FRL1* in frogs (Kinoshita et al. 1995), *one-eyed pinhead* (*oep*) in zebrafish (Zhang et al. 1998), and *Cfc* in chick (Colas and Schoenwolf 2000; Schlange et al. 2001). These proteins contain an epidermal growth factor (EGF)-like domain and a cripto-FRL1-Cryptic (CFC) domain (Adamson et al. 2002). A number of genetic and biochemical studies have shown that EGF–CFC members are absolutely essential for signaling by Nodal and the Vg1/GDF1 subset of TGFβ superfamily ligands, but not for signaling by Activin (Cheng et al. 2003; Saloman et al. 2000; Schier 2003; Yeo and Whitman 2001). For example, in zebrafish, the total absence of *oep* (maternal and zygotic) gives rise to the same phenotype as the double mutant *cyclops;squint* (Gritsman et al. 1999). Mouse Cfc1 has been shown to bind to Acvr1b and it seems that in its absence Nodal cannot bind to the receptor complex (Fig. 15; Sakuma et al. 2002; Yan et al. 1999; Yeo and Whitman 2001).

In the mouse, *Cfc1* is expressed in the node and midline, and bilaterally symmetric in the LPM, from the head fold to the 6–8-somite stage (Shen et al. 1997).

Its pattern of expression in the chick is very similar (Fig. 9C; Schlange et al. 2001). Consistent with Cfc1 mediating signaling by Nodal, the phenotype exhibited by *Cfc1* mutants, characterized by the absence of left-sided gene expressions and right isomerism, is quite similar to the hypomorphic or conditional *Nodal* mutants (Brennan et al. 2002; Gaio et al. 1999; Norris et al. 2002; Yan et al. 1999).

In addition to coreceptors, Nodal signaling is also modulated by the presence of extracellular antagonists that add an extra level of complexity to the system (Shi and Massague 2003). Lefty and Cerberus proteins are well-known Nodal antagonists (Fig. 15).

Lefty genes include *Lefty1* and *Lefty2* in the mouse and zebrafish, *LEFTYB* and *EBAF* in humans and *Lefty1* in the chick (Ishimaru et al. 2000; Kosaki et al. 1999a; Meno et al. 1997; Rodriguez-Esteban et al. 1999; Thisse and Thisse 1999). A number of experiments have demonstrated that Lefty proteins are capable of blocking Nodal signaling and recently they have also been shown to block GDF1/Vg1 signaling (Bisgrove et al. 1999; Cheng et al. 2004; Meno et al. 1999). Accordingly, the phenotypes of loss and gain of function of *Lefty* are respectively explained by enhanced or reduced Nodal signaling both in zebrafish and in the mouse (Meno et al. 1999; Meno et al. 1998; Meno et al. 2001; Thisse and Thisse 1999).

Recently, Lefty proteins have been shown to function by blocking EGF–CFC coreceptors, therefore preventing their interaction with the receptor complex and hence Nodal signaling (Sakuma et al. 2002). Since Lefty also blocks Gdf1 signaling, the situation is quite intricate. In the mouse *Lefty2* is expressed in the LPM in a pattern very similar to that of *Nodal* while *Lefty1* is expressed in the midline in a pattern very similar to that of *Gdf1*. In the absence of *Lefty2* in the left LPM (Meno et al. 2001), the expression of *Nodal* is expanded both in terms of time and space. Since Lefty directly interacts with Nodal and with the co-receptor EGF–CFC, a possible explanation for the *Lefty2* phenotype is that Lefty binds Nodal and limits its propagation as well as preventing the activation of Nodal's signaling receptor (Branford and Yost 2002; Chen and Schier 2002; Meno et al. 1999; Meno et al. 2001).

As described, both *Nodal* and *Lefty2* have left-side specific enhancers (ASE) that contain essential binding sites for the transcriptional factor Foxh1 and Smad2 (Saijoh et al. 2000; Schier and Shen 2000). This suggests an autoregulatory feedback mechanism whereby Nodal induces and maintains its own expression, while activating Lefty2. Once secreted Lefty2 antagonizes Nodal, as described above, limiting its propagation and duration of expression in the LPM (Hamada et al. 2002).

3.3
Bmp Signaling and Left–Right Asymmetry

Bone morphogenetic proteins (Bmps) were discovered by their ability to induce bone when applied subcutaneously (Kingsley 1994). Later it was shown that Bmps played very important roles during embryonic development, including the specification of the primary body axes in *Drosophila* and *Xenopus* (Hogan 1996; Mishina 2003).

Bmps are secreted proteins belonging to the TGFβ superfamily (Massague 1996; Shi and Massague 2003). Like other members of the TGFβ family, Bmps signal through receptors with serine/threonine kinase activity (Shi and Massague 2003). The ligand assembles with one dimer of type I receptor and another of type II receptor to form the receptor complex (Table 5). Three R-Smads, Smad1, 5 and 8, have been shown to be specific for Bmp signaling and to become phosphorylated by receptor type I upon ligand binding (Shi and Massague 2003).

As mentioned earlier, several studies indicate that Bmp signaling plays a fundamental role in LR development although it may vary depending on the stage and may not be conserved in different species.

Several Bmps are expressed in the gastrulating chick embryo (Rodriguez-Esteban et al. 1999; Yokouchi et al. 1999). *Bmp2*, *Bmp4*, and *Bmp7* are expressed in the primitive streak, midline, and peripheral LPM (Fig. 9D). Of these only *Bmp4* shows a transient LR asymmetry in expression on the right side of Hensen's node during stages 5 to 7HH (Fig. 8B; Monsoro-Burq and Le Douarin 2001). The pattern of expression of *Bmp4* in the node is complementary to that of *Shh* and is believed to be responsible for *Shh* restriction of expression to the left side of the node at stage 5HH. Exogenous application of Bmp to the left side of Hensen's node abolishes *Shh* expression and therefore the left-sided pathway (Monsoro-Burq and Le Douarin 2001; Piedra and Ros 2002). Thus, during the early phase of chick LR development, Bmp signaling is an essential component of the right-sided pathway, as occurs in *Xenopus*.

In *Xenopus*, *Vg1*, the ortholog of mouse *Gdf1*, is considered to be a left-side determinant (Branford et al. 2000; Hyatt and Yost 1998; Ramsdell and Yost 1999). Independently of its uniform left–right expression in the blastula embryo, misexpression experiments have shown that *Vg1* expression directs the left-side pathway leading to *Nodal* expression in the left LPM, since its ectopic expression in the right randomizes LR asymmetries (Hyatt et al. 1996). In contrast, Bmp signaling, mediated by the receptor Acvr1 (Alk2), controls the right-side pathway and antagonizes the Vg1 left-sided pathway. Based on all these data, Hyatt and Yost (1998) proposed the left–right coordinator model in which two different and antagonizing pathways control the establishment of LR asymmetries, with Bmp signaling constituting an important component of the right-sided pathway.

However, at early somite stages Bmp signaling acts as a positive regulator of Nodal expression facilitating the left-sided pathway both in the chick and in mice (Fujiwara et al. 2002; Piedra and Ros 2002; Schlange et al. 2002). It was previously thought that Bmp from the peripheral LPM prevented *Nodal* expression in its left LPM domain (Rodriguez-Esteban et al. 1999; Yokouchi et al. 1999). This idea was based on studies with the chick gene *Car*, which is transiently expressed in the left LPM in a pattern reminiscent of but not similar to that of Nodal (Fig. 13A; Sect. 2.3). Car is a member of the Dan/Cerberus family of Bmp antagonists capable of binding Nodal and BMPs. Therefore, *Car* was thought to be required in order to block endogenous Bmp signaling in the left LPM as a pre-requisite for *Nodal* initiation of expression. However, careful experiments performed in early somite stage embryos showed that Bmp signaling consistently and rapidly induced *Nodal* as well as *Car* expression (Piedra and Ros 2002; Schlange et al. 2002). Other genes with symmetric or right-sided expression such as *Cfc* and *Snail1* were also positively regulated by Bmp signaling (Piedra and Ros 2002; Schlange et al. 2002; Schlange et al. 2001). Bmp signaling is also known to be detrimental for midline development where it suppresses *Lefty1* expression (Piedra and Ros 2002; Yokouchi et al. 1999). However, the positive regulation of *Nodal* expression caused by ectopic Bmp application was not mediated by impairment of the midline, as demonstrated by barrier experiments (Piedra and Ros 2002). Since Bmp signaling rapidly and consistently induces *Nodal*, as well as *Cfc*, it has been suggested that its positive action on *Nodal* expression may be mediated by Cfc (Piedra and Ros 2002; Schlange et al. 2002). Thus, in sum, the role Car may play in the chick left LPM is presently unknown and loss-of-function experiments would be required to help unravel its function. In the chick, the current evidence indicates that Bmp signaling may acts as a right-side determinant at early stages by controlling restriction of Shh signaling to the left. Later it functions as a left-side determinant by facilitating *Nodal* expression in the LPM.

It seems likely that *Car* is the chick ortholog of *Cerberus* (*Cer*) in the mouse. However, *Cer* does not replicate the *Car* asymmetric pattern of expression in the left LPM. Recently another related gene, *Dand5*, has been shown to be expressed on the right side of the node, where it is thought to block spread of Nodal signaling to the right (Marques et al. 2004; Pearce et al. 1999).

Further evidence for the implication of Bmp signaling in the left-sided pathway came from the study of *Bmp4* mutants. In the early mouse embryo *Bmp4* is first expressed in the extraembryonic mesoderm and later in the posterior primitive streak and bilaterally in the LPM (Fujiwara et al. 2002). *Bmp2* is also expressed bilaterally in the LPM while *Bmp7* is expressed in the node and midline (Solloway and Robertson 1999). *Bmp4* expression in the extraembryonic ectoderm is required for the normal morphological development of the node and primitive streak, as demonstrated by the *Bmp4* mutant embryo, which exhibits a posterior bulge projecting ventrally and an abnormally flat node which, however, presents monocilia in most cases (Fujiwara et al. 2002). In these mutants *Nodal* expression is patchy in the node and absent in the LPM where *Lefty2* is predominantly absent and *Cfc1* not maintained. The embryos only survive up to the 20-somite stage, so that heart looping is the only morphological asymmetry that can be analyzed. In the absence of *Bmp4*, more than half of the embryos lack evidence of heart looping resulting in mesocardia (Fujiwara et al. 2002). The study of tetraploid chimeric embryos, in which the embryo is only made of *Bmp4* null cells while the extraembryonic tissue is made of wild type tetraploid cells that cannot contribute to the embryo but otherwise are normal, showed that extraembryonic Bmp4 was sufficient to restore the morphology of the node and primitive streak but not the alterations in gene expression and heart looping.

However, the phenotype of *Bmp4* null tetraploid chimeric embryos is opposite regarding *Nodal* expression to that of the *Smad5* mutant and *Acvr1* null chimeras.

Mice mutant for *Smad5*, one of the Bmp-specific Smads, show bilateral expression of *Nodal* in the LPM in half of the mutant embryos, as well as defects in heart looping and embryo turning (Chang et al. 2000; Winnier et al. 1995). The *Acvr1* null chimeric embryos (Table 5) show a similar phenotype with bilateral expression of left-side markers (Kishigami et al. 2004). These results support the notion that Bmp signaling may normally repress *Nodal* in the right LPM, although, since *Smad5* is expressed ubiquitously, it is unclear how this process could be achieved. *Smad5* mutants also show decreased or absent *Lefty1* in the midline, also suggesting that bilateral *Nodal* expression is a consequence of the midline defect. The Acvr1 is also used by other pathways besides Bmps, therefore making difficult to determine which specific pathway is altered in this mutation.

Interestingly, the node and its derivatives express proteins that are antagonists of the Wnt and Bmp pathways thus blocking signaling to the receptors of these pathways. *Noggin* and *Chordin* respectively encode two antagonists of Bmp signaling, both of which are expressed in the node and notochord of mouse and chick embryos (Bachiller et al. 2000; Connolly et al. 1997; Streit et al. 1998). Neither *chordin* null mice nor *noggin* null mice show alterations in laterality (Bachiller et al. 2003; McMahon et al. 1998). However, the double mutants, besides lacking normal forebrain development, also exhibit randomization of *situs* (Bachiller et al. 2000). The laterality defects may be interpreted as caused by the absence of midline exhibited by the double mutant.

In summary, all the data mentioned above seem to indicate that the role played by Bmp signaling in LR patterning is complex and probably differs depending on the stage. The discrepant phenotypes observed after disrupting distinct components of the Bmp signaling pathway may also reflect distinct functions of different BMPs at specific developmental stages. Current knowledge is compatible with Bmp signaling positively regulating the left-sided molecular pathway, particularly *Nodal* expression, in the early-somite avian and mammal embryo. At earlier stages Bmp4 participates in proximodistal patterning of the epiblast and morphogenesis of the node in the mouse whereas it controls asymmetric gene expression in Hensen's node in the chick.

4
Mouse Models of Laterality

Animal models of disease are powerful tools for analyzing and dissecting the physiopathology of defects at the genetic, molecular, cellular, and embryological levels. Therefore, those spontaneous or experimentally induced mutations that cause laterality defects offer an invaluable opportunity to better understand the etiopathology of the disease as well as the molecular components of the LR pathways. Nevertheless, it is important to stress that mice may present specificities, mainly due to the particularities of the development of the early rodent embryo, and these may not be replicated by human embryos. The list of mouse mutations affecting LR development is continuously increasing, particularly by the recent addition of mutations affecting genes with a role in ciliary biogenesis and function. Several recent reviews have included comprehensive tables (Hamada et al. 2002; Levin 2005; Maclean and Dunwoodie 2004) as we have also done in Table 6. Next we will consider those mutations that have proved most relevant for the field.

The best-known mouse mutation affecting LR development is the autosomal recessive *inversus viscerum* (*iv*) mutation, also the first to be discovered (Hummel and Chapman 1959). The mutation arose spontaneously and was identified because stomachs of some of the newborn mice were on the right side of the abdomen. Mutant homozygous mice are viable and fertile and show randomization of the organ *situs*: 50% showing *situs solitus* and 50% showing *situs inversus* (Brueckner et al. 1989; Layton 1976; McGrath et al. 1992). Besides the randomization of *situs*, the *iv* mutation has an increased incidence of associated heart malformations (Icardo and Sanchez de Vega 1991). The mutation causes randomization of expression of the left-side markers *Nodal*, *Lefty2*, and *Pitx2*, that show all possible patterns of expression, left, right, bilateral, or absent (Collignon et al. 1996; Lowe et al. 1996; Meno et al. 1996; Piedra et al. 1998; Ryan et al. 1998).

The *iv* mutation was mapped to the distal arm of mouse chromosome 12 and was later identified as a glutamate to lysine substitution in the highly conserved motor domain of the outer arm axonemal dynein *Dnahc11* (also *lrd*) gene that results in a loss-of-function mutation (Supp et al. 1997). The engineered, targeted disruption of *Dnahc11* gives the *iv* phenotype (Supp et al. 1999). *Dnahc11* is expressed in the monociliated nodal cells during embryonic development and in a subset of the adult ciliated epithelia such as the oviduct and choroid plexus, but not in other tissues such as testis or kidney (McGrath and Brueckner 2003; McGrath et al. 2003; Supp et al. 1999). *Dnahc11* is shown to localize to cilia and, as already mentioned, in the node is only found in the central motile monocilia while the peripheral immotile monocilia lack *Dnahc11* expression (McGrath and Brueckner 2003). The *iv* mutation renders the nodal monocilia immotile supporting the conclusion that the *iv* phenotype is caused by the loss of nodal-flow (Nonaka et al. 1998). The normal absence of *Dnahc11* expression in respiratory epithelia and in the testis of adults explains the absence of respiratory disease or infertility in the *iv*/*iv* homozygous mice (McGrath and Brueckner 2003).

Legless (*lgl*) is an insertional mutation that causes a deletion of more than 600 kb from the distal arm of chromosome 12 that includes the *Dnahc11* gene (Supp et al. 1999; Supp et al. 1997). *Lgl* mutants exhibit randomization of *situs* similarly to the *iv* mutation as well as limb malformations and other phenotypic traits due to the deletion of other genes (Schneider et al. 1999). Very interestingly, mutations in the human homolog *DNAH11* have been found in human patients with laterality defects, indicating a high degree of functional conservation in vertebrates (Bartoloni et al. 2002).

Table 6 Mouse genes whose mutations give a LR phenotype

Gene symbol	Name; role	Expression site	Model	Phenotype	Heart situs	*Nodal* expression in node	LPM expressions	Midline expressions	Nodal cilia	References
Mutations of Nodal pathway components										
Nodal	TGF-β family signaling molecule	Node, stronger on the left side and left LPM	Hypomorphic (*Nodal*$^{Neo/Neo}$)	• Right isomerism of the heart, lungs, and spleen • Reversed position of great arteries • Random rotation of stomach	Random	Reduced or absent	*Nodal*, *Pitx2*, and *Lefty2* absent	*Lefty1* absent *Shh* normal		Saijoh et al. 2003
			Left-specific deletion ASE (*Nodal*$^{\Delta600/\Delta600}$)	• Partial right isomerism	Ventrally posi-tioned	Loss of asym-metry	*Nodal* and *Lefty2* reduced, *Pitx2* reduced anteriorly	*Lefty1* and *Shh* absent		Norris et al. 2002
			Node-specific deletion (*Nodal*$^{\Delta/\Delta}$)	• Right isomerism		Markedly attenu-ated	*Nodal* normal	*Lefty1* absent		Brennan et al. 2002

Table 6 (continued)

Gene symbol	Name; role	Expression site	Model	Phenotype	Heart situs	*Nodal* expression in node	LPM expressions	Midline expressions	Nodal cilia	References
			Node-specific deletion/ Null (*Nodal* $^{\Delta/-}$)	• Right isomerism	Random	Absent	*Nodal*, *Pitx2*, and *Lefty2* absent	*Lefty1* absent *Shh* normal		
			Nodal $^{+/-}$/ *Foxa2* $^{+/-}$		Positioning defect		*Nodal* bilateral			Collignon et al. 1996
Acvr2b	Activin receptor II b; Nodal receptor	Epiblast and ectoderm	Knock-out	• Right isomerism • Cardiac anomalies • Hypoplasia of the spleen						Oh and Li 1997; Oh and Li 2002
			ActRIIb $^{-/-}$/*iv* $^{-/-}$	• Right isomerism						
			ActRIIb $^{-/-}$/ *Nodal* $^{+/-}$	• Right isomerism						
			ActRIIb $^{-/-}$/ *ActRIIA* $^{+/-}$	• Right isomerism						

Table 6 (continued)

Gene symbol	Name; role	Expression site	Model	Phenotype	Heart situs	*Nodal* expression in node	LPM expressions	Midline expressions	Nodal cilia	References
Cfc1 (Cryptic)	Cryptic family 1; Nodal coreceptor	Node, notochord, prospective floor plate, and LPM	Knock-out	• Pulmonary right isomerism • Randomization of embryo turning and abdominal situs • Vascular heterotaxia	50% dextrocardia or mesocardia	Normal	*Nodal*, *Pitx2*, and *Lefty2* absent	*Lefty1* absent *Shh* normal		Gaio et al. 1999; Yan et al. 1999
Furin (Pace/ Spc1)	Paired basic amino acid cleaving enzyme; involved in Nodal protein processing	Extraembryonic tissues, node, primitive heart, and LPM	Knock-out	• Defects in ventral closure	Cardiac bifida	Normal	*Nodal* normal, *Pitx2*, and *Lefty2* upregulated or bilateral	*Lefty1* normal		Roebroek et al. 1998; Constam and Robertson 2000b

Table 6 (continued)

Gene symbol	Name; role	Expression site	Model	Phenotype	Heart situs	*Nodal* expression in node	LPM expressions	Midline expressions	Nodal cilia	References
Psck6 (Spc4/ Pace4)	Proprotein convertase subtilisin/ kexin type 6; involved in Nodal protein processing	Extraembryonic tissues, foregut endoderm.	Knock-out	• Left isomerism • Anterior central nervous system defects • Severe cardiac malformations		Normal	*Nodal*, *Pitx2*, and *Lefty2* Bilateral	*Lefty1* normal		Constam and Robertson 2000a
Lefty2 (Ebaf)	Left–right determination factor 2; TGF-β family (Nodal antagonist)	Left LPM	Left-specific deletion (*Lefty2*$^{ASE\ -/-}$)	• Left isomerism	Normal	Normal	*Nodal* and *Pitx2* bilateral (posterior level) and prolonged	*Lefty1* normal and prolonged		Meno et al. 2001; Sakuma et al. 2002
Lefty1 (Leftb)	Left–right determination factor 1; TGF-β family (Nodal antagonist)	Left prospective floor plate	Knock-out	• Thoracic Left isomerism	Normal		*Nodal* and *Lefty2* bilateral *Pitx2* bilateral (anterior level)	*Shh* normal		Meno et al. 1998; Perea-Gomez et al. 2002

Table 6 (continued)

Gene symbol	Name; role	Expression site	Model	Phenotype	Heart situs	*Nodal* expression in node	LPM expressions	Midline expressions	Nodal cilia	References
Dand5 (Cerl-2, Dte)	DAN domain family member 5; BMP and Nodal antagonist	Node, stronger on the right side	Knock-out	• Variety of laterality defects (left pulmonary isomerism, thoracic situs inversus)	Leftward or ventral heart looping	Normal	*Nodal*, *Pitx2*, and *Lefty2* bilateral, left or right	*Lefty1* and *Shh* normal		Marques et al. 2004
Smad2	MAD homolog 2; TGF-β intracellular effector	Ubiquitous	Epiblast-specific deletion ($Smad2^{CA}$)	• Anterior central nervous system defect • Absent of functional midline barrier			*Nodal* bilateral	*Shh* absent		Vincent et al. 2003; Waldrip et al. 1998
			$Smad2^{+/-}$ $Nodal^{+/-}$	• Right pulmonary isomerism • Opposite turning • Cardiac defect	Abnormal heart looping					Nomura and Li 1998

Table 6 (continued)

Gene symbol	Name; role	Expression site	Model	Phenotype	Heart situs	*Nodal* expression in node	LPM expressions	Midline expressions	Nodal cilia	References
Foxh1 (*Fast1/2*)	Forkhead activin signal transducer 2; transcriptional partner of Smad proteins	LPM and heart	Knock-out	• Absence of node and midline structures (variable severity)	No survival		*Pitx2* absent *Lefty2* downregulated	*Lefty1* and *Shh* absent		Hoodless et al. 2001; Yamamoto et al. 2001
			Conditional deletion in the LPM (*Foxh1*$^{c/-}$)	• Right isomerism	50% inverted or mesocardia	Normal	*Nodal*, *Pitx2*, and *Lefty2* absent			Yamamoto et al. 2003
Foxa2 (*HNF3-β*)	Hepatocyte nuclear factor 3-*β*	Node, notochord, floorplate, and gut	Tetraploid chimeras	• Absence of node and notochord • Defects in dorsal-ventral patterning of the neural tube	Absence of heart looping		*Nodal* absent *Lefty2* absent or bilateral	*Lefty1* and *Shh* absent		Dufort et al. 1998; Ang and Rossant 1994; Weinstein et al. 1994

Table 6 (continued)

Gene symbol	Name; role	Expression site	Model	Phenotype	Heart situs	*Nodal* expression in node	LPM expressions	Midline expressions	Nodal cilia	References
Pitx2	Paired-like homeodomain transcription factor 2	Cephalic meso-derm and left LPM	Knock-out	• Right pulmonary isomerism • Failure of ventral body wall closure • Septal and valve defects, single atrium	Rightward looping					Kitamura et al. 1999; Lin et al. 1999; Lu et al. 1999; Gage et al. 1999
			Isoform-specific deletion (*Pitx2c*$^{-/-}$)	• Right atrial and pulmonary isomerism • Vascular defects	Rightward looping					Bamforth et al. 2004; Liu et al. 2002
			Hypomorpho (*Pitx2*$^{neo/neo}$)	• Right pulmonary isomerism • Failure of ventral body wall closure • Septal and valve defects	Rightward looping					Gage et al. 1999

Table 6 (continued)

Gene symbol	Name; role	Expression site	Model	Phenotype	Heart situs	*Nodal* expression in node	LPM expressions	Midline expressions	Nodal cilia	References
Cited2	Transcriptional repression of HIF1; nuclear transactivator	Ventral node, cardiac crescent, anterior lateral mesoderm, and paraxial mesoderm	Knock-out	• Right thoracic isomerism • Incomplete penetrance of abdominal isomerism	Random	Normal Reduced (Bamforth et al. 2004)	*Nodal*, *Pitx2* and *Lefty2* absent	*Lefty1* absent (anterior level) *Shh* normal		Bamforth et al. 2004; Weninger et al. 2005
Mutations of Sonic hedgehog pathway components										
Shh	Sonic hedgehog; signaling molecule	Node, midline mesoderm, and gut endoderm	Knock-out	• Left pulmonary isomerism • Randomization of axial turning • Cardiac defects • Midline defects	Normal		*Nodal* and *Lefty2* left or bilateral *Pitx2* bilateral	*Lefty1* absent		Meyers and Martin 1999; Chiang et al. 1996; Izraeli et al. 1999; Tsukui et al. 1999

Table 6 (continued)

Gene symbol	Name; role	Expression site	Model	Phenotype	Heart situs	*Nodal* expression in node	LPM expressions	Midline expressions	Nodal cilia	References
			Double mutant $Shh^{-/-}/Ihh^{-/-}$	• Failure of embryonic turning • Small, linear heart tube • Open gut	Absence of heart looping		*Pitx2* absent			Zhang et al. 2001
Smo	Smoothened homolog; involved in hedgehog signaling	Ubiquitous	Knock-out	• Failure of embryonic turning • Small, linear heart tube • Open gut	Absence of heart looping	Variable and of random asymmetry	*Nodal*, *Pitx2*, and *Lefty2* absent	*Lefty1* absent		Zhang et al. 2001
Disp1 (*DispA*)	Dispatched homolog 1; involved in release Hh signal	Ubiquitous	Knock-out	• Inflated pericardial sac • Abnormal trunk morphology • Failure of embryonic turning	Defective heart looping	Variable	*Nodal* absent	*Shh* and *Foxa2* normal in notochord		Ma et al. 2002

Table 6 (continued)

Gene symbol	Name; role	Expression site	Model	Phenotype	Heart situs	*Nodal* expression in node	LPM expressions	Midline expressions	Nodal cilia	References
Sufu	Suppressor of fused homolog; negative modulator of Hh signaling	Central nervous system	Knock-out	• Failure to undergo embryonic turning • Abnormal somites • Open neural tube	Random	Expanded in and around the node	*Pitx2* absent or bilateral	*Shh* and *Foxa2* up-regulated	Present (abnormal node architecture)	Cooper et al. 2005
Stil (*Sil*)	Scl/Tal1 interrupting locus	Ubiquitous	Knock-out	• Axial midline defects • Delay or failure of neural tube closure	Random		*Nodal*, *Pitx2*, and *Lefty2* bilateral	*Lefty1* and *Shh* absent		Izraeli et al. 2001; Izraeli et al. 1999
Mutations of BMP pathway components										
Bmp4	Bone morphogenetic protein 4; TGF-β family member	Posterior primitive streak and LPM	Knock-out	• Abnormal node morphology (flat or slightly convex)	Mesocardia (defective heart looping)	Patchy or reduced	*Nodal*, *Lefty2*, and *Cfc1* absent	*Shh* normal	Present	Fujiwara et al. 2002

Table 6 (continued)

Gene symbol	Name; role	Expression site	Model	Phenotype	Heart situs	*Nodal* expression in node	LPM expressions	Midline expressions	Nodal cilia	References
				• Absence of allantois						
			Tetraploid chimeras	• Normal node	Mesocardia	Reduced	*Nodal* reduced	*Shh* normal		
Acvr1 (*Alk2*)	Activin receptor I; BMP receptor	Node and midline, weakly in LPM	Chimeras	• Failure or reverse turning	Defective heart looping	Symmetric or stronger on the right	*Nodal*, *Pitx2*, and *Lefty2* bilateral	*Lefty1* absent *Shh* normal		Kishigami et al. 2004
Smad5	MAD homolog 5; TGF-β intracellular effector	Ubiquitous	Knock-out	• Defect in heart looping and embryonic turning	Random or defective	Normal	*Nodal* and *Pitx2* bilateral *Lefty2* low level bilateral	*Lefty1* absent		Chang et al. 2000
Chrd/Nog	Chordin/ Noggin; BMP antagonists	Node and axial mesoderm	Double mutant $Chrd^{-/-}$/ $Nog^{-/-}$	• Prosen-cephalon defects	Random			*Shh* and *Foxa2* normal (absent in rostral midline)		Bachiller et al. 2000

Table 6 (continued)

Gene symbol	Name; role	Expression site	Model	Phenotype	Heart situs	*Nodal* expression in node	LPM expressions	Midline expressions	Nodal cilia	References
Mutations of Notch pathway components										
Notch1 *Notch2*	Notch gene homolog 1/2	Notch1 in presomitic mesoderm Notch2 in node, notochord, and neural groove	Double mutant *Notch1*$^{-/-}$/ *Notch2*$^{neo/neo}$	• Defects in axial rotation	Random					Krebs et al. 2003
Dll1	Delta-like1; DSL ligand of Notch pathway	Primitive streak and posterior mesoderm	Knock-out	• Randomization of embryo turning	Random	Absent	*Nodal* and *Lefty2* absent *Pitx2* left, absent or bilateral	*Lefty1* absent *Shh* and *Foxa2* normal	Present and functional	Raya et al. 2003; Krebs et al. 2003; Bettenhausen et al. 1995
				• Randomization of embryo turning		Absent or of random asymmetry	*Nodal*, *Pitx2*, and *Lefty2* random		Abnormal cells intermixed with ciliated cells	Przemeck et al. 2003

Table 6 (continued)

Gene symbol	Name; role	Expression site	Model	Phenotype	Heart situs	*Nodal* expression in node	LPM expressions	Midline expressions	Nodal cilia	References
Mutations of other ligands										
Fgf8	Fibroblast growth factor 8	Posterior primitive streak and presomitic mesoderm	Hypomorphic ($Fgf8^{neo/neo}$)	• Right isomerism (50%)	Inverted or meso-cardia	Low level or absent	*Nodal*, *Pitx2*, and *Lefty2* absent			Meyers and Martin 1999
Gdf1	Growth differentiation factor 1; TGF-β family member	Node, midline and LPM	Knock-out	• Right pulmonary isomerism • Visceral *situs inversus* • Cardiac anomalies	Random	Normal	*Nodal*, *Pitx2* and *Lefty2* absent	*Lefty1* absent		Wall et al. 2000; Rankin et al. 2000
Wnt3a	Wingless-related MMTV integration site 3A	Primitive streak and dorsal posterior node	Knock-out	• Variety of laterality defects (Right pulmonary isomerism, *situs* ambiguus)	Random or meso-cardia	Spatially reduced	*Nodal*, *Pitx2*, and *Lefty2* delayed, then bilateral in posterior LPM	*Lefty1* absent *Shh* normal	Present (PC1 down-regulated)	Nakaya et al. 2005

Table 6 (continued)

Gene symbol	Name; role	Expression site	Model	Phenotype	Heart situs	*Nodal* expression in node	LPM expressions	Midline expressions	Nodal cilia	References
Mutations of transcription factors										
T	Brachyury; Transcription factor	Primitive streak, mesoderm, head process, and notochord	Knock-out	• Lack of a mature notochord • Dorsalization of neural tube • Cardiac defect	Random	Absent	*Nodal* and *Lefty2* absent *Pitx2* bilateral	*Lefty1* absent (only in a broad patch of ventral cells)		King et al. 1998; Tsang et al. 1999
Bapx1 *(Nkx3.2)*	Bagpipe homeobox gene 1 homolog; transcription factor	Stronger on right lateral plate mesoderm, somites	Knock-out	• Lethal; perinatal skeletal dysplasia and asplenia				*Shh* normal		Schneider et al. 1999; Tribioli and Lufkin 1999

Table 6 (continued)

Gene symbol	Name; role	Expression site	Model	Phenotype	Heart situs	*Nodal* expression in node	LPM expressions	Midline expressions	Nodal cilia	References
Mutations of genes implicated in ciliary biogenesis and function										
Kif3a	Kinesin family member 3A; anterograde transport in cilia	Ubiquitous	Knock-out	• Randomization of embryo turning	Random		*Ptix2* and *Lefty2* bilateral	*Shh* absent	Absent or extremely short	Takeda et al. 1999; Marszalek et al. 1999
Kif3b	Kinesin family member 3B; anterograde transport in cilia	Node	Knock-out	• Randomization of embryo turning and abdominal situs	Random		*Lefty2* bilateral or absent		Absent (only basal bodies)	Nonaka et al. 1998
Dync2h1 (*Dnchc2*)	Dynein cytoplasmic 2 heavy chain 1; retrograde transport in cilia	Node	Knock-out and chemically induced	• Abnormal brain morphology • Polydactyly	Random heart looping		*Nodal* bilateral	*Shh* normal	Present but short or bulged (Epithelium: immotile)	Huangfu and Anderson 2005
Dnahc5 (*Mdnah5*)	Dynein, axonemal, heavy chain 5; retrograde transport in cilia	Node	Insertional mutation	• Situs inversus • Respiratory infections • Hydrocephalus	Random				Epithelium: Immotile (Loss of axonemal outer arms)	Ibanez-Tallon et al. 2002; Olbrich et al. 2002

Table 6 (continued)

Gene symbol	Name; role	Expression site	Model	Phenotype	Heart situs	*Nodal* expression in node	LPM expressions	Midline expressions	Nodal cilia	References
Dnahc11 (iv,lrd)	Dynein, axonemal, heavy chain 11 (inversus viscerum, left–right-dynein); retrograde transport in cilia	Node, ciliated epithelium	Knock-out Inversus viscerum (*iv*, spontaneous mutation)	• Randomization of situs • Heterotaxia	Random	Asymmetry concordant with expression in LPM	*Nodal* and *Pitx2* left, right, bilateral, or absent *Lefty2* left or right	*Lefty1* left or right	Present but immotile (Epithelium: motile)	Supp et al. 1999; Supp et al. 1997
			Legless (*lgl*, insertional mutation)	• Randomization of embryo turning and abdominal situs • Hindlimb, brain, and craniofacial malformations						

Table 6 (continued)

Gene symbol	Name; role	Expression site	Model	Phenotype	Heart situs	*Nodal* expression in node	LPM expressions	Midline expressions	Nodal cilia	References
Dync2li1 (*D2lic*)	Dynein cytoplasmic 2 light intermediate chain 1; retrograde transport in cilia	Node	Knock-out	• Failure or randomization of embryo turning	Random	Reduced or symmetric	*Nodal* *Pitx2* and *Lefty2* left, right, bilateral, or absent	*Lefty1* absent *Shh*, *T* and *Foxa2* reduced	Absent or stunted	Rana et al. 2004
Rfx3	Rfx family member 3; regulates *D2lic* expression	Node	Knock-out	• Left pulmonary isomerism • Cardiac malformations	Inverted or mesocardia		*Nodal* and *Lefty2* bilateral, weaker in right side	*Lefty1* and *Foxa2* normal	Present but stunted	Bonnafe et al. 2004
Ift88 (*Polaris*, *Tg737*, *orpk*)	Intraflagellar transport protein 88	Node	Chemically induced and Knock-out *Oak Ridge Polycystic Kidney* (*orpk*, insertional mutation, hypomorph)	• Laterality defects • Neural tube defects • Renal and pancreatic cysts • Polydactyly • Hydrocephalus	Random	Normal	*Nodal* and *Lefty2* bilateral	*Shh* reduced *Foxa2* absent	Absent	Murcia et al. 2000; Moyer et al. 1994; Huangfu et al. 2003

Table 6 (continued)

Gene symbol	Name; role	Expression site	Model	Phenotype	Heart situs	*Nodal* expression in node	LPM expressions	Midline expressions	Nodal cilia	References
Ift172 (Wim)	Intraflagellar transport protein 172	Node	Chemically induced	• Abnormal brain morphology	Random		*Nodal* bilateral	*Shh* normal	Absent	Huangfu et al. 2003
Invs (Inv)	Inversin	In 9+0 cilia	*Inversion of embryonic turning* (*inv*, insertional mutation)	• Situs inversus 90%	Inverted cardiac loop	Stronger on the right	*Nodal*, *Pitx2* and *Lefty2* right	*Lefty1* normal	Present and motile (very weak leftward flow)	Okada et al. 1999; Yokoyama et al. 1993; McQuinn et al. 2001; Morgan et al. 1998; Watanabe et al. 2003
Pkd2	Polycystic kidney disease 2; PC2	Ubiquitous (Localizes to nodal monocilia)	Knock-out	• Right isomerism • Randomization of embryo turning • Renal cyst • Cardiac defects	Random	Normal	*Nodal* and *Lefty2* absent *Pitx2* Bilateral (posteriorly)	*Lefty1* absent *Shh* normal	Present	Barr et al. 2001; Wu et al. 2000; Pennekamp et al. 2002

Table 6 (continued)

Gene symbol	Name; role	Expression site	Model	Phenotype	Heart situs	*Nodal* expression in node	LPM expressions	Midline expressions	Nodal cilia	References
Foxj1 (*Hfh4*)	Hepatocyte nuclear factor forkhead homologue 4; transcription factor	Node, ciliated epithelium	Knock-out	• Randomization of situs • Cardiovascular malformations	Random				Present in node but absent in ciliated epithelium	Brody et al. 2000; Chen et al. 1998
Other mutations										
Aldh1a2 (*Raldh2*)	Aldehyde dehydrogenase family 1 subfamily 2	Posterior mesoderm during gastrulation	Knock-out	• Defects in anteroposterior body axis • Hindbrain	No obvious LR asymmetry	Normal	*Nodal*, *Pitx2* and *Lefty2* normal	*Lefty1* normal		Niederreither et al. 2001

Table 6 (continued)

Gene symbol	Name; role	Expression site	Model	Phenotype	Heart situs	*Nodal* expression in node	LPM expressions	Midline expressions	Nodal cilia	References
Fltn	Flectin; extra-cellular matrix molecule	Right side predominance at the outflow tract, left side predominance at the ventricular portion of the tubular heart	Immuno histochemistry							Tsuda et al. 1998
Fts (fused toes)			Spontaneous mutation	• Fusion of fore limb digits and a thymic hyperplasia	Random		*Nodal* and *Lefty2* bilateral			Heymer et al. 1997
Mgat1	Mannoside acetylglucosaminyl-transferase 1	Ubiquitous	Knock-out	• Situs inversus • Neural tube defects	Random					Metzler et al. 1994

Table 6 (continued)

Gene symbol	Name; role	Expression site	Model	Phenotype	Heart situs	*Nodal* expression in node	LPM expressions	Midline expressions	Nodal cilia	References
nt (No turning)			Spontaneous mutation	• Failure of embryonic turning • Neural tube and notochordal defect	Random		*Nodal* and *Lefty2* left, right, bilateral, or absent	*Lefty1* absent *Shh* abnormal		Melloy et al. 1998
Ofd1	Oral facial digital syndrome 1		Knock-out	• Failure of embryonic turning • Polydactyly	Random and mesocardia		*Nodal* and *Pitx2* Bilateral	*Shh* normal in notochord but absent in floorplate	Absent	Ferrante et al. 2001; Ferrante et al. 2006
Poll (Pol-λ)	DNA polymerase X family member λ		Knock-out	• Situs inversus • Hydrocephalus • Chronic sinusitis • Male infertility	Random				Respiratory cilia immotile (defect of inner dynein arms)	Kobayashi et al. 2002

Table 6 (continued)

Gene symbol	Name; role	Expression site	Model	Phenotype	Heart situs	*Nodal* expression in node	LPM expressions	Midline expressions	Nodal cilia	References
Rttn	Rotatin; novel trans-membrane protein	Notochord and meso-derm	Insertional mutation	• Failure of embry-onic turning • Neural tube defects	Random	Normal	*Nodal* bilateral *Pitx2* and *Lefty2* bilateral or absent	*Shh* absent anteriorly		Faisst et al. 2002
Slc8a (NCX-1)	solute carrier family 8 (sodium/ calcium exchanger), member 1	Right lateral meso-derm and developing heart	Knock-out	• Lack of a heartbeat	Normal					Linask et al. 2001; Koushik et al. 2001
Zic3	Zinc-finger protein of the cerebellum 3	Central nervous system, tailbud, and somites	Knock-out	• Failure or abnormal embryonic turning • Neural tube defects	Abnormal position	Initiated but not maintained, no asym-metry	*Nodal* and *Pitx2* random			Purandare et al. 2002

A unique situation is seen in the insertional mutation called *inversion of embryonic turning* (*inv*), which results in *situs inversus* in the majority of the homozygous embryos (McQuinn et al. 2001; Yokoyama et al. 1993). The homozygous mice are not viable and die during the first week after birth, probably due to the associated kidney and liver disease (cysts). They show a consistent reversal in the direction of the LR polarity with a common pattern of *situs inversus*. Accordingly the expression of the left-side genes *Nodal*, *Pitx2*, and *Lefty2* is reversed in the LPM, as is the direction of heart looping (Lowe et al. 1996; Meno et al. 1996; Ryan et al. 1998). As previously mentioned, full *situs inversus* is also obtained in *Xenopus* after misexpression of the active form of Vg1 (homolog of Gdf1) on the right side (Hyatt et al. 1996; Hyatt and Yost 1998). In a sense, the fact that the *inv* mutation reverses the *situs* of every homozygotic individual is reminiscent of the *sinistral* mutation in snails (Gurdon 2005).

The *inv* mutation was caused by the insertion of the minigene tyrosinase in the mouse chromosome 4. The insertion caused an important deletion in a novel gene called *Inversin* (*Invs*) that codes for a protein containing 15 consecutive ankyrin repeats at its amino terminus (Mochizuki et al. 1998; Morgan et al. 1998; Nurnberger et al. 2006). The insertion of the minigene caused a deletion of exons 4–12 of the *inv* gene resulting in a complete loss of function of the gene. The *inv* has orthologs in several species including chick, *Xenopus*, zebrafish, and humans (Morgan et al. 2002). Comparative sequence analysis revealed that the ankyrin repeat domain is highly conserved as well as the two IQ domain characteristic of calcium-independent calmodulin binding. Calmodulin is a calcium sensor involved in multiple Ca^{2+}-dependent cellular processes.

Very importantly, in humans *INVERSIN* has been identified as the gene responsible for nephronophthisis type 2 (NPHP2; Otto et al. 2003). NPHP2 is an autosomal recessive cystic kidney disease that leads to end-stage renal failure in children and is also accompanied by multiple extrarenal manifestations including pancreatic islet dysplasia, anomalies of the hepatobiliary system, and *situs inversus*. The presence of *situs inversus* may be linked to the functional severity of the mutation in humans.

Inversin is expressed from the two-cell stage of embryonic development (Eley et al. 2004). At the subcellular level, inversin is expressed in the 9+0 cilia of the node, renal tubules, and retina, but not in the 9+2 cilia of the trachea, oviducts, or ependyma (Watanabe et al. 2003). Inversin expression continues in the adult in a variety of tissues such as the renal tubules, the hepatobiliary ducts, and the intestinal tract. Expression appears to affect all nodal cells, including the central cells with motile cilia and the peripheral cells with immotile sensory cilia. In the absence of inversin the nodal cilia move but the flow they generate is slow and turbulent (Okada et al. 1999; Yokoyama et al. 1993). This result is difficult to explain with either the nodal-flow or the two-cilia hypothesis.

A great advance has been made recently with the identification of inversin's function as a molecular switch between the canonical and non-canonical Wnt signaling cascades (Simons et al. 2005). Equally important is the finding that fluid flow can increase levels of inversin in ciliated cells. However, despite the growing information on inversin function, the exact manner in which inversin regulates *Nodal* expression remains unknown.

In addition to *inversin* recent investigations have revealed that other genes associated with renal cystic disease are important for LR development (Igarashi and Somlo 2002). Probably the most important is *Pkd2* the gene responsible for the autosomal dominant polycystic kidney disease type 2, that affects children. *Pkd2* encodes the polycystin2 (PC2) protein that is an integral membrane protein that functions as a Ca^{2+}-selective channel. PC2 physically interacts with polycystin1 (PC1), also an integral membrane protein implicated in autosomal polycystic kidney disease. Targeted disruption of *Pkd2* in the mouse results in dominant polycystic kidney disease in heterozygosis, while in homozygosis the condition is more severe (Pennekamp et al. 2002). The PC1-PC2 complex localizes to the primary cilia of renal epithelium where it is thought to function as the molecular sensor for fluid flow (Praetorius and Spring 2001). In response to cilia bending due to fluid shear stress, PC2 serves as a specific mechanosensitive Ca^{2+} channel that allows Ca^{2+} ions to enter the cell and transduce the mechanic stimulus. Very interestingly, *Pkd2* mutants develop *situs inversus* (Pennekamp et al. 2002). Since it has been shown that PC2 functions as a molecular mechanosensor in the kidney, it was suggested that it could play the same function in the node as assigned by proposals. While *Dnahc11* is only expressed in the motile central monocilia, PC2 is expressed, together with *Inversin*, in all nodal monocilia including peripheral non-motile cilia. The latter observations were the basis of the previously described two-cilia hypothesis (Sect. 2.1.2; McGrath et al. 2003; Qiu et al. 2005; Tabin and Vogan 2003).

Another set of mutants with important laterality defects are mutations affecting proteins involved in ciliogenesis or cilia function. Two of these mutations affect genes involved in the intraflagellar transport proteins, *Ift88* and *Ift172*, which are known to assemble in the same IFT complex. The absence of either *Ift88* or *Ift172* results in absence of cilia in the node, and bilateral expression of left-sided markers in the LPM (Huangfu et al. 2003). Again, these observations stress the relationship between nodal cilia and establishment of LR patterning. Remarkably, it has recently been shown that these two ciliary proteins, as well as the kinesin component Kif3a and the dynein Dnahc2 also participate in mediating hedgehog signaling in vertebrates (Huangfu and Anderson 2005). The *Ift88* and the *Ift172* mutations show phenotypes characteristic of defective Shh signaling such as the absence of ventral neural cell types in the neural tube. Genetic studies have demonstrated that Ift88 and Ift172 are required at a step downstream of Ptc (the receptor of Shh) and upstream of Gli3 (Huangfu et al. 2003). Therefore, in addition to the highly evolutionary conserved role of IFT, kinesin and dynein proteins in ciliogenesis, it seems that in vertebrates they are also involved in transduction of hedgehog signaling genes (see also Sect. 3.1).

A large subset of mutations with LR phenotype are those that cause defective morphogenesis or function of the midline. For the most part, these mutations have been considered in Sect. 2.4 and they mostly yield phenotypes of bilateral expression of left-side markers. However, a variety of LR phenotypes are possible depending on the time of onset and extent of the midline defect (Izraeli et al. 1999).

Finally, we will mention the insertional mutation called *Fused toes* (*Ft*), which is manifested, in homozygosis, through randomization of *situs*, severe brain alterations, and bilateral expression of left-sided genes (Heymer et al. 1997; van der Hoeven et al. 1994; Volkmann et al. 1996). The transgene integration in *Ft* resulted in a deletion of 1.6 Mb on mouse chromosome 8, including the entire IroquoisB gene cluster (Peters et al. 2002). The possible implication of the Iroquois gene cluster in the phenotype remains to be determined. Interestingly, heterozygous $ft^{+/-}$ mutant mice present fusion of the forelimb digits and thymus hyperplasia.

5 Genetics of Human Alterations of Organ *Situs*

As previously mentioned, alterations in organ *situs*, particularly *situs ambiguus*, are well-recognized in human pathology and exhibit complex and varied presentations that are difficult to categorize. Most commonly *heterotaxia* and *isomerism* associate with congenital heart disease (CHD), which may exist as the only expression of the laterality defect. Most cases of *heterotaxia* are sporadic although many familial cases have also been documented. *Heterotaxia* is a consistent feature in several autosomal recessive syndromes and has also been shown to associate with diabetic embryopathy and occasionally with conditions involving karyotypic rearrangement (Kuehl and Loffredo 2002; Lin et al. 2000; Zhu et al. 2006).

The genetics of the majority of heterotaxias, either syndromic or isolated, are presently unknown. An added difficulty is that diagnosis of heterotaxy syndromes is daunting not only because thorough examinations are required but also because many of the phenotypic signs are common to several syndromes. Indeed, some syndromes are really heterogeneous conditions in which disorders of very different etiology cluster together. Recently, studies performed in model organisms, and previously described in this review, have unraveled, at least in part, the genetic pathways that govern the establishment of asymmetries in the LR axis and have provided an ample number of candidate genes for analysis in human malformations. It is not known how many of the hundred or so genes involved in LR development in mice have conserved their function in humans, but the data available indicate that the pathways may be very similar. Indeed, the most fruitful approach to date for the identification of mutations responsible for human laterality syndromes has been to search for mutation in genes known to play this role in animal models, particularly in mice. In any case, the numerous mice models of laterality together with the broad phenotype spectrum seen in humans predict a highly complex underlying genetic network that is only beginning to emerge.

The hypothesis that mutations in the human homologs of mouse genes involved in LR patterning would lie at the root of human *heterotaxia*, stimulated the identification and structural analysis of the corresponding human genes and subsequent screening for mutations in individuals with *situs* abnormalities. Numerous molecular studies, many of which used the single-strand conformation polymorphism (SSCP), were undertaken in the search for polymorphisms or mutations in candidate human genes. A series of genes identified in these studies as responsible for human laterality defects that are included in Table 7.

One of the earliest reports on the genetics of human heterotaxia identified point mutations in the cytoplasmic tail of the gap junction alpha 1 (*GJA1*, also *CONNEXIN43;* OMIM 121014) protein in children with heart malformations and laterality defects (Britz-Cunningham et al. 1995). This report stimulated research that allowed identification of the influence of gap junctions in LR patterning (Levin and Mercola 1998b; Levin and Mercola 1999). However, the involvement of *GJA1* remains uncertain since later studies have been unable to detect *GJA1* mutations in patients with *heterotaxia* (Debrus et al. 1997; Gebbia et al. 1996; Splitt and Goodship 1997). Targeted mutation of *Connexin43* in the mouse is neither associated with laterality defects (Reaume et al. 1995) although the mutants exhibit abnormal heart looping, which has been attributed to a defect in the migration of the heart neural crest (Ya et al. 1998). Gap-junction communication appears to play an important role in the regulation of neural crest migration and development and therefore in modulating mammalian cardiac development (Dasgupta et al. 2001). The fact that about 20 connexins have been identified in the human and mouse genomes and that most cells express several connexins (Evans and Martin 2002) may hamper the ability of assigning a role for gap junctions in establishing the LR axis due to redundancy or compensation among different connexins.

As expected, a mutation in human *NODAL* in heterozygosis has been shown to cause *situs ambiguus* (Gebbia et al. 1997). But the essential function of Nodal in gastrulation probably makes homozygotic mutations nonviable.

Also, mutations in the human gene *ACVR2B*, (OMIM 602730), the homolog of the *Acvr2b* in mice, have been found in patients with typical LR malformations indicating that the pathways are in fact conserved (Kosaki et al. 1999b). However, these mutations are extremely rare and are only found in a very small percentage of affected individuals (see Table 7).

Two recent reports have identified heterozygous mutations in the human *CFC1* (OMIM 605194) gene in patients with LR laterality defects and in patients with isolated transposition of the great arteries (TGA) and double-outlet right ventricle (DORV) (Bamford et al. 2000; Goldmuntz et al. 2002). These studies are of great interest as they demonstrate a common genetic etiology for heterotaxy syndromes and isolated cases of TGA and DORV, indicating that these conditions may represent different degrees of the same genetic condition.

As in the mouse, two lefty genes have been identified in humans, designated as *LEFTYA* and *LEFTYB* (Kosaki et al. 1999a). These two genes localize in chromosome 1 (1q42) in a region syntenic to 1H5 in mouse, the region where the mouse lefty genes are located (Meno et al. 1997). *LEFTYA* was previously identified as endometrial-bleeding-associated factor (*EBAF*; Kothapalli et al. 1997) and the name *EBAF* has been selected for the human nomenclature system (OMIM

Table 7 Human genes involved in heterotaxy

Gene	Product/role	OMIM	Locus	Mutations/series	Allelic variants		Reference
ACVR2B	Activin receptor II b	602730	3p22-p21.3	3/112 sporadic 0/14 familial	Amino acid substitution R40H Amino acid substitution V494I	Left–right axis malformations	Kosaki et al. 1999b
CFC1	Member of EGF-CFC gene	605194	2q21.1	4/144	Amino acid substitution R112C 1 bp deletion in exon 6 (522C) Amino acid substitution R189C	Heterotaxy visceral autosomal	Bamford et al. 2000
					CFC1:1bp deletion (522C)	Heterotaxy	
					Nodal mutation		
				1/64	Duplication of 20 bp	Dextro looped transposition of the great arteries	Goldmuntz et al. 2002
				1/22	1 bp deletion in exon 6	Double outlet right ventricle	
CRELD1	Cysteine-rich protein with EGF-LIKE domain 1	607170	3p25.3		Amino acid substitution R329C Amino acid substitution T311I Amino acid substitution P162A	Susceptible to atrioventricular septal defect	Robinson et al. 2003 Zatyka et al. 2005

Table 7 (continued)

Gene	Product/role	OMIM	Locus	Mutations/series	Allelic variants		Reference
				1/11	Amino acid substitution R107H	Susceptible to atrioventricular septal defect with heterotaxy syndrome	Robinson et al. 2003
DNAI 1	Dynein axonemal intermediate chain type 1	604366	9p13-p21		1 bp insertion in intron 4 bp insertion in codon 95	Ciliary dyskinesia primary 1	Guichard et al. 2001; Pennarun et al. 1999
					Amino acid substitution G515S Deletion of 12 bp	Kartagener syndrome	
DNAH5	Dynein axonemal heavy chain type 5	603335	5p15-p14		1 bp insertion in exon 34 Amino acid substitution G3519R 1 bp deletion Amino acid substitution Q610X 1 bp insertion	Kartagener syndrome Ciliary dyskinesia primary 3	Olbrich et al. 2002
DNAH11/ LRD	Dynein axonemal heavy chain type 11	603339	Chr. 7		Amino acid substitution R2852X	Situs inversus viscerum	Bartoloni et al. 2002

Table 7 (continued)

Gene	Product/role	OMIM	Locus	Mutations/series	Allelic variants		Reference
GJA1/ CON-NEXIN 43	Gap junction protein 43KD	121014	6q21-q23.2		Amino acid substitution R362Q Amino acid substitution R376Q	Hypoplastic left heart syndrome (included atrioventricular septal defect)	Britz-Cunningham et al. 1995; Dasgupta et al. 2001
INVS	Inversin	243305	9q31	7/7 families	Amino acid substitution R603X Amino acid substitution L493S	Infantile nephronophthisis (NPHP2)	Otto et al. 2003; Gagnadoux et al. 1989
LEFTYA/ EBAF	TGF-β-family signaling molecule (Left–right determination factor 2)	601877	1q42.1	2/112	Amino acid substitution R314X Amino acid substitution S342K	Left–right axis malformations	Kosaki et al. 1999a
NKX2.5	NK2 transcription factor related	600584	5q34	One family	7 bp deletion	Atrial septal and atrioventricular conduction defect	Watanabe et al. 2002

Table 7 (continued)

Gene	Product/role	OMIM	Locus	Mutations/series	Allelic variants		Reference
NODAL	Member of TGFβ transcription factor	601265	Chr. 10	One case	Amino acid substitution R183Q	Situs ambiguus	Gebbia et al. 1997
PITX2	Paired like homeodoma in transcription factor	691542	4q25-q26		Amino acid substitution L54G G to C transversion Amino acid substitution T68P A to G transition Amino acid substitution R91P Amino acid substitution W193X Amino acid substitution V45L	Rieger syndrome type I	Semina et al. 1996
					21 bp duplication		Priston et al. 2001
THRAP2/ PROSIT 240	Thyroid hormone receptor associated protein 2	608771	12q24	3/97	Amino acid substitution E251G Amino acid substitution R1872H Amino acid substitution D2023G	Dextro-looped transposition of the great arteries	Muncke et al. 2003

Table 7 (continued)

Gene	Product/role	OMIM	Locus	Mutations/series	Allelic variants		Reference
TTTC8/ BBS8	Tetratricope ptide repeat domain 8	608132	14q32.1	3/120	6 bp deletion 3 bp deletion	Bardet-Bield syndrome	Ansley et al. 2003
ZIC3	Zinc finger protein of cerebellum 3	300265	Xq26.2	9 families 3 sporadic/>200 cases	Amino acid substitution T325 M Amino acid substitution C270X	Heterotaxy visceral X link	Gebbia et al. 1997
					Amino acid substitution Q249X Amino acid substitution C253S Amino acid substitution K405E		Ware et al. 2004
					Transversion 1741A-T (from Lys to stop) Amino acid substitution P217A	Transposition of great arteries X linked Congenital heart disease, non-heterotaxy	Megarbane et al. 2000

601877). Screening for mutations in LEFTY genes in patients with *situs* alterations yielded two mutations in *EBAF* as possible causes of the disease. These mutations were a nonsense and a missense mutation and the two patients showed a similar phenotype including left pulmonary isomerism and cardiac defects reminiscent of the phenotype of *Lefty1* mutant mice (Kosaki et al. 1999a; Meno et al. 1998).

Nkx2.5 is a gene essential for cardiac development in mice (Lyons et al. 1995). It has also been implicated in maintenance of *Pitx2* expression once activated by Nodal (Shiratori et al. 2001). Interestingly, a deletion in *CSX* (OMIM 600584), the human homolog of *Nkx2.5* in mouse, has been documented in a family with atrial septal defect and atrioventricular node conduction failure (Watanabe et al. 2002). One member of this family also exhibited *heterotaxia* suggesting a possible involvement of *CSX* in laterality defects, perhaps through interaction with *Pitx2* (Shiratori et al. 2001).

ZIC3 (OMIM 300265) is a zinc finger transcription factor that has been identified as the gene mutated in cases of X-linked *heterotaxia* and in patients with isolated heart disease (Gebbia et al. 1997; Purandare et al. 2002; Ware et al. 2004). Interestingly, a role for Zic3 has been also identified in the mouse and in *Xenopus* (Kitaguchi et al. 2000; Purandare et al. 2002) possibly by mediating the establishment of the left-side pathways.

Finally, it is worth mentioning that mutations in *SHH* and *PITX2* humans genes have not been related to laterality defects. *SHH* is responsible in humans for the autosomal dominant holoprosencephaly type 3 (Belloni et al. 1996) and *PITX2* for the autosomal dominant Axenfeld–Rieger (OMIM 180500) syndrome characterized by dental hypoplasia, ocular anterior chamber anomalies, and umbilical anomalies (Rieger 1935; Semina et al. 1996). In marked contrast to what occurs in the mouse, the haploinsufficiency of these two genes in humans is sufficient to give a strong phenotype. These are not the only cases since most genes implicated in LR patterning, as mentioned above, give rise to a phenotype in heterozygosis whereas mice heterozygous for mutations are phenotypically normal. In some cases the explanation may reside in the fact that naturally occurring mutations in man may act as dominant negative, or yield a product of unknown function, whereas mutations in mice are engineered to result in a complete loss of function.

5.1 The Primary Ciliary Dyskinesia Syndrome

One in about 20,000 individuals presents a complete reversal of organ *situs*—*situs inversus* or *situs inversus totalis*. The phenotype is readily identifiable and usually produces no symptoms of disease, confirming the theoretical consideration that the reversed mirror image arrangement of organs is as good as *situs solitus*. *Situs inversus* has been known to exist for centuries (see for example McMannus 2002) usually being a chance finding in postmortem studies. Nowadays it is commonly diagnosed by prenatal ultrasound and, unless it carries associated malformations, does not require treatment.

Between 20 and 25% of people with *situs inversus* have an underlying condition called primary ciliary dyskinesia (PCD; OMIM 242650), a rare genetic disorder caused by defective ciliary and flagellar function (Van's Gravesande and Omran 2005). It is a phenotypically heterogeneous condition with several loci and genes identified as responsible. The predominant inheritance pattern is autosomal recessive but occasional cases of autosomal dominant and X-linked inheritance have been documented. Half of the patients with PCD also display *situs inversus*, this association being known as Kartagener's syndrome (OMIM 244400), a subtype of PCD. Kartagener's syndrome is characterized by the triad of chronic sinus infections, bronchiectasis due to increased mucous secretions from the lungs, and situs inversus (Afzelius 1976).

PCD is caused by the immotility or dysmotility of cilia and flagella. Therefore, the symptoms are conspicuous in those organs whose adequate function requires ciliary function, as in the respiratory tract, or flagellar function, as for spermatozoid motility. Typical ultrastructural defects consist of total or partial absence of dynein arms (about 80%), dislocation of central tubules (about 10%), radial spokes defects (about 6%), and peripheral microtubule anomalies (Ibanez-Tallon et al. 2002). The most common defect is loss of the outer dynein arms affecting almost all cilia (Afzelius 1985). Dyneins are high molecular weight molecular motors associated with microtubules showing ATPase activity and responsible for ciliary movement (see Appendix). Less frequent are other ciliary anomalies found in PCD: ciliary aplasia, alterations of the basal apparatus, hockey stick cilia, and long cilia (Ibanez-Tallon et al. 2002).

Given the high functional complexity and the multiple components of cilia and flagella, PCD would be expected to show considerable genetic heterogeneity and indeed genetic studies performed in families with PCD have found multiple loci to be implicated in its pathogenesis. So far the strategies adopted to identify genes responsible for PCD have recognized mutations in three genes encoding dynein chains.

The first discovery was based on the recognition that mutations in the *IC78* gene of *Chlamydomonas Reihardtii* (a biflagellate alga used for studies of cilia) render the alga immotile due to defective cilia that lacked the outer dynein arms, in a structural pattern similar to that documented in human PCD (Perrone et al. 1998). Taking advantage of the evolutionary conservation of genes encoding axonemal proteins, Pennarun et al. (1999) concentrated on the isolation of the human ortholog of *IC78*, designated *DNAI1* (*dynein axonemal intermediate chain-1*, OMIM 604366) and demonstrated loss-of-function mutations in a patient with PCD. The success of this study demonstrated the value of search for candidate-gene approaches and this was confirmed and extended by later studies (Guichard et al. 2001).

Mutations in two other genes encoding dyneins have been identified in a small percentage of PCD/Kartagener syndrome individuals, but it is to be expected that new components will be identified in the future. One of these is the *DNAHC11* (*dynein axonemal heavy chain-11*, OMIM 603339) a complex gene that comprises 82 exons (Bartoloni et al. 2002). Interestingly this gene is the human ortholog of the mouse *Dnahc11* (also *left–right dynein, lrd*) that was isolated as the gene responsible for the spontaneous mouse mutation *iv* (*inversus viscerum*), the first and best-known mouse model of *situs inversus*. Mutations in *DNAH11* cause *situs inversus totalis* reinforcing the analogy between the molecular control of *situs* in mice and humans (Bartoloni et al. 2002). The third dynein involved in LR asymmetry in humans is the *DNAH5* (*dynein axonemal heavy chain-5*, OMIM 603335), whose mutations cause PCD with randomization of *situs* (Olbrich et al. 2002). The murine ortholog of *DNAH5* is expressed in the node in a pattern similar to *dnah11*, and patients with mutations in *DNAH5* have immotile cilia (visualized in the respiratory tract) that at the ultrastructural level lack the outer dynein arms (Olbrich et al. 2002).

The enormous efforts made in this field in recent years have also revealed that other dynein components can be ruled out as candidates for a role in PCD. For example, the screening for *DNAL1* (*dynein axonemal light chain-1*) mutations in individuals with PCD did not identify any sequence variation, thus eliminating its involvement in the etiology of PCD (Horvath et al. 2005). The murine *Dnal1* ortholog shows a pattern of expression similar to that of *Dnah5* and the two proteins have been shown to physically interact.

Several other candidates genes found not responsible for PCD include *DNAH9* (*dynein heavy chain 9*), *DNAI2* (*dynein intermediate chain 2*), and *LC8* and *TCTE3* (Bartoloni et al. 2001; Neesen et al. 2002; Pennarun et al. 2000). The human homolog of the transcription factor Foxj1, which regulates *Dnah11* expression, has also been ruled out as a gene implicated in the etiopathology of PCD (Maiti et al. 2000).

Interestingly, the *DNAH7* (*dynein heavy chain 7*) is a component of the inner dynein arm that is synthesized but not assembled in a case of PCD, indicating that the defect is in another gene not yet identified (Zhang et al. 2002).

Although mutations in each particular ciliary component are involved in only a small minority of patients with PDC, given the high number of genes coding for components of cilia or flagella, it may be that the sum of their mutations represents a considerable proportion of affected individuals. In this context, (Pazour et al. 2006) have performed a systematic *in silico* search to identify the human homologs of proteins associated with the *Chlamydomonas reinhardtii* outer dynein arm, which is the most well characterized outer arm of any species. Their results show that most of the outer dynein subunits in *Chlamydomonas* have at least one ortholog in humans, and have identified a total of 24 human genes as potentially involved in outer arm assembly and therefore good candidates for analysis in PCD patients.

Another protein with ciliary subcellular localization is *INVS* (*INVERSIN*, OMIM 243305), which, as already mentioned (Sect. 4), is involved in nephronophthisis 2 (NPHP2), an autosomal recessive cystic kidney disorder in children (Otto et al. 2003). A role for inversin in LR axis specification in humans has been confirmed by the finding of a patient with complete *situs inversus* and homozygous truncating mutation in inversin (Otto et al. 2003).

6
Concluding Remarks

The last 10 years have seen remarkable advances in our understanding of how the internal asymmetries typical of the vertebrate body are established and controlled. The use of different development models has permitted investigators to uncover fascinating ways of creating asymmetry, such as using the activity of the nodal cilia. A next step now is to define the extent of conservation across species of the mechanisms that have been discovered. Another is to refine the ultimate way in which they control the asymmetric morphogenesis of the internal organs.

A host of studies has also unraveled the involvement of many genes in the left-right patterning pathway. Based on this knowledge the genetic basis of human laterality defects are beginning to be revealed. A major challenge now is to understand how all these genes control left-right development as well as the complex set of interactions established among them.

Here we have tried to present in an orderly fashion and discuss our current knowledge of the development of the left-right axis in vertebrates, particularly mammals. Several aspects require further clarification but with the rapid advance in the field we will soon have a more complete picture of how left–right asymmetries emerges in the vertebrate body.

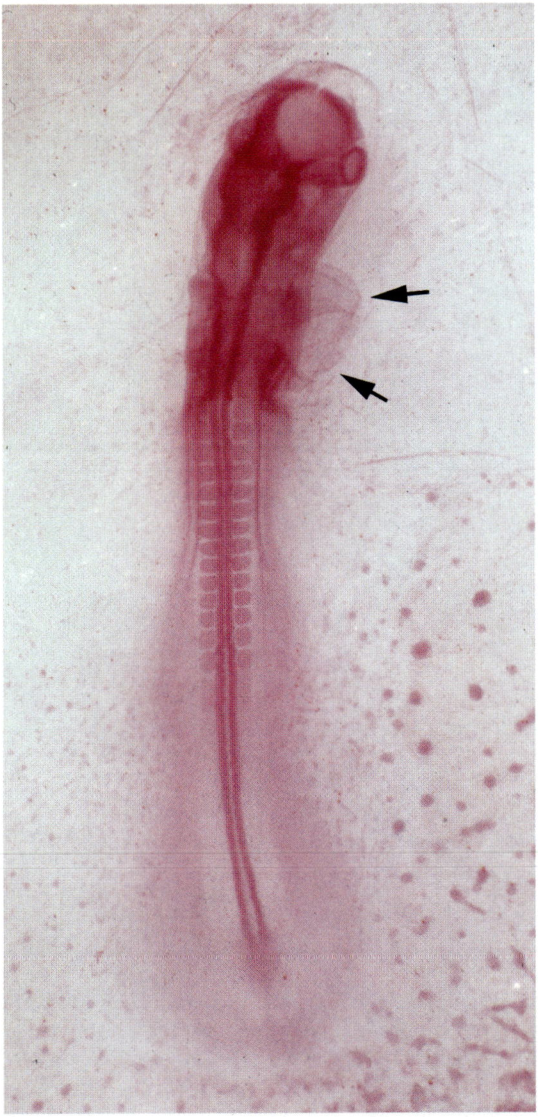

Fig. 1 The first morphological asymmetry is the formation of the heart loop. Dorsal view of a 12HH (2 days of incubation) chick embryo stained with hematoxylin showing the right heart loop (*arrows*)

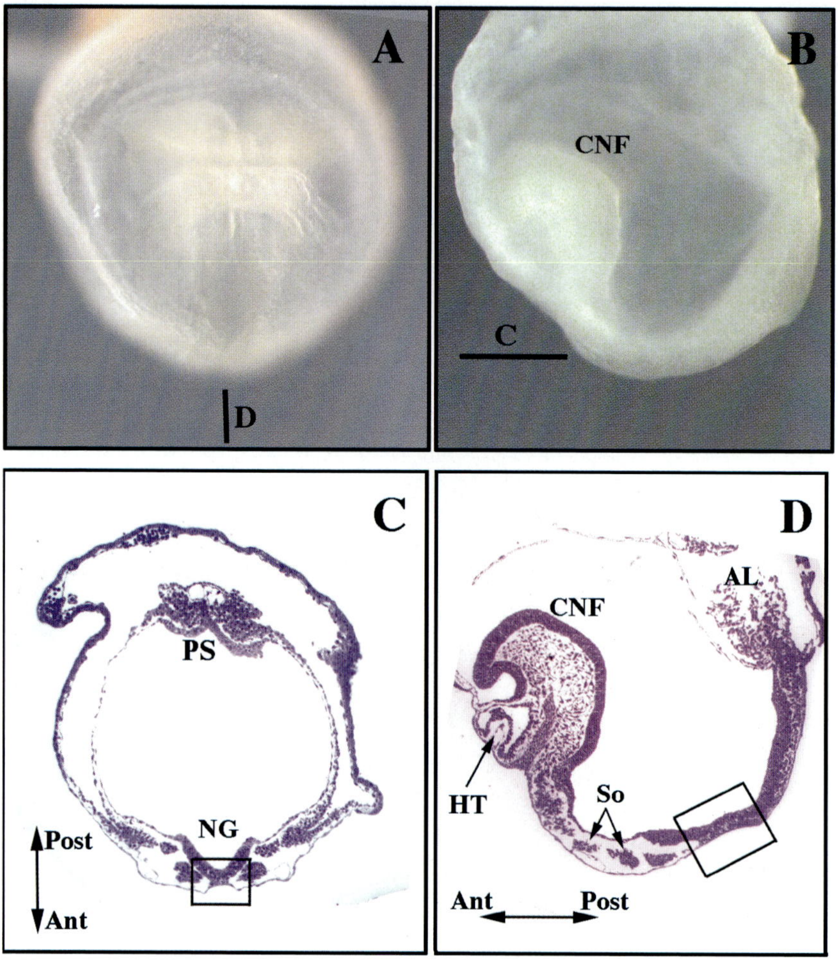

Fig. 2A–D Morphology of the mouse embryo at embryonic day 8.5. **A** Anterior and **B** lateral views of a 3–4-somite embryo immediately after fixation. **C** Transverse semi-thin section of the embryo in **B** as indicated by the *C* line. **D** Sagittal semi-thin section of the embryo in **A** as indicated by the *D* line. The *boxed areas* mark the areas depicted in Fig. 4. *PS*, primitive streak; *NG*, neural groove; *CNF*, cephalic neural fold; *HT*, heart tube; *So*, somites; *AL*, allantois

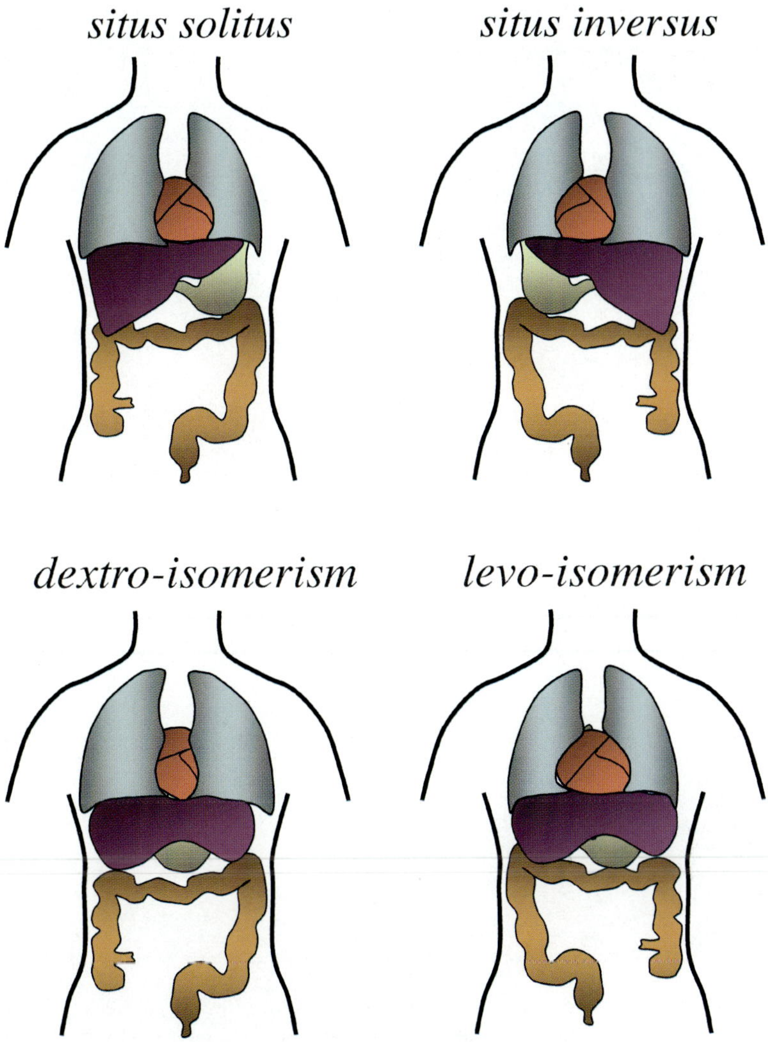

Fig. 3 Schematic diagram illustrating four possible patterns of visceral organ arrangement. Normal left–right development results in *situs solitus*. Mirror-image reversal of all the organs results in *situs inversus*. Bilateral multi-lobed lungs and midline liver are typical of *dextro-isomerism* while lungs with lesser lobes are typical of *levo-isomerism*

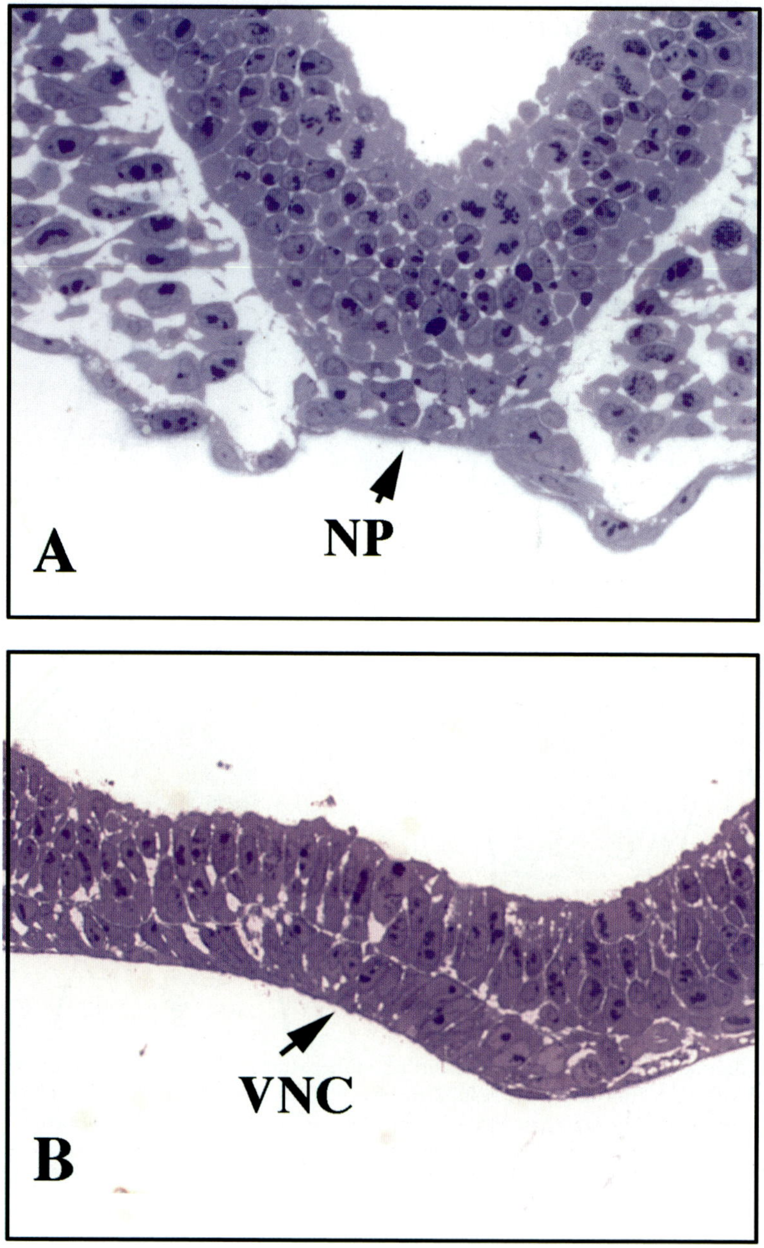

Fig. 4A, B Morphology of the mouse notochordal plate and node. **A** Semi-thin transverse section showing the morphology of the notochordal plate (magnification of the *boxed area* in Fig. 2C). **B** Semi-thin sagittal section through the node showing its two-layered structure (magnification of the *boxed area* in Fig. 2D). *NP*, notochordal plate; *VNC*, ventral nodal cells

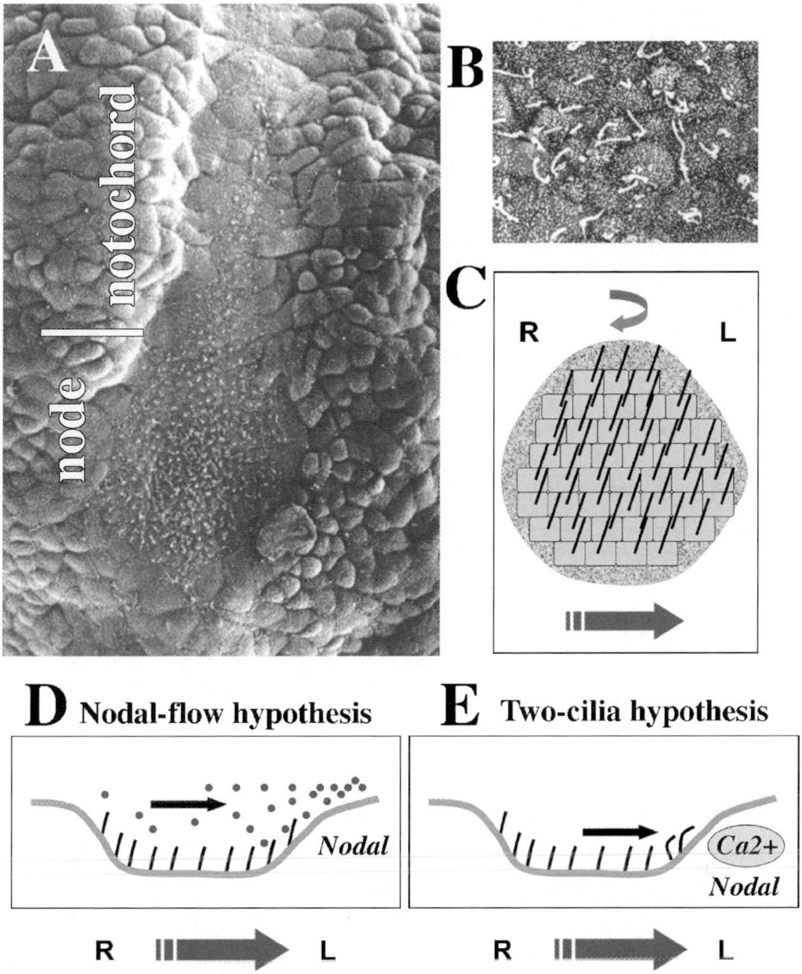

Fig. 5A–E Morphology of the mouse node. **A** Scanning electron micrograph showing the ventral aspect of the node of a E8.5 mouse embryo. The node and notochordal plate regions are indicated. **B** Detailed view of the ventral nodal cells, each of them bearing a monocilium. **C** Schematic diagram illustrating the triangular shape of the mouse node, the direction of the ciliary rotational movement, and the direction of the nodal-flow. **D** and **E** are diagrams illustrating the "nodal-flow" and the "two-cilia" hypotheses, respectively. The *small circles* in **D** represent the morphogen particles and the *arrow* the direction of the nodal flow. See text for details. *R*, right; *L*, left

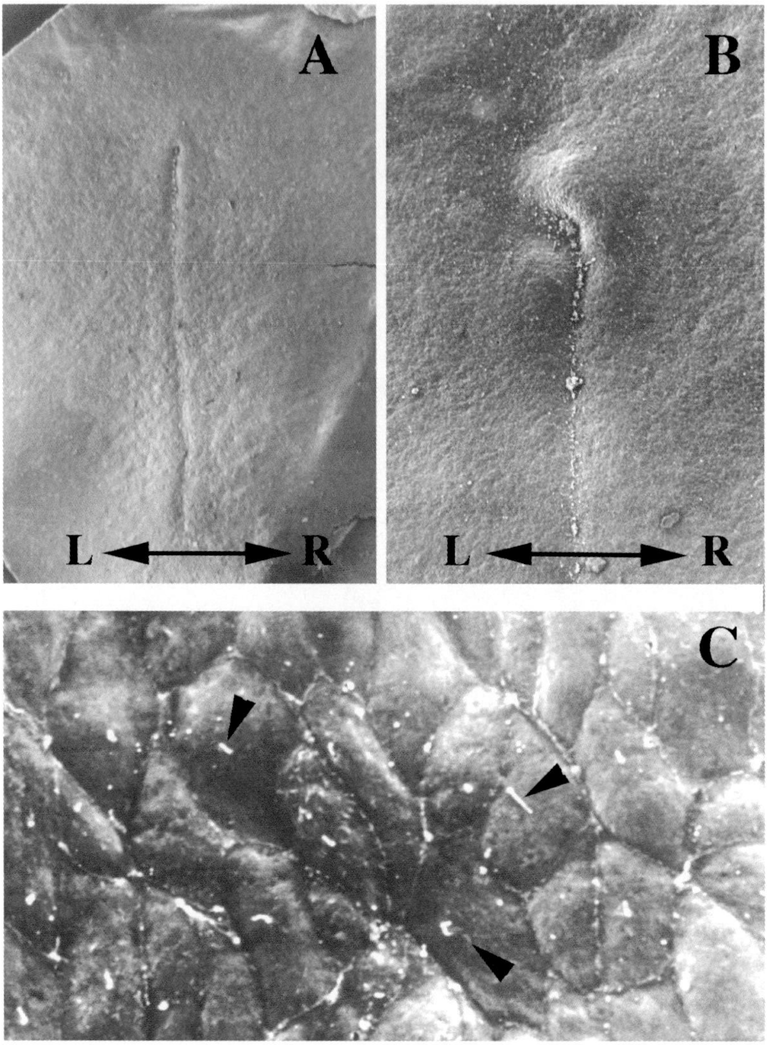

Fig. 6A–C Morphology of the chick Hensen's node. **A** Scanning electron micrograph of a 4HH chick embryo (dorsal view). **B** Scanning electron micrograph showing the dorsal aspect of a 5HH Hensen's node that at this stage is clearly asymmetric. Note the presence of cell debris in the groove of the primitive streak. **C** Detailed view of dorsal epiblast cells some of them bearing a small monocilia (*arrowheads*). *R*, right; *L*, left

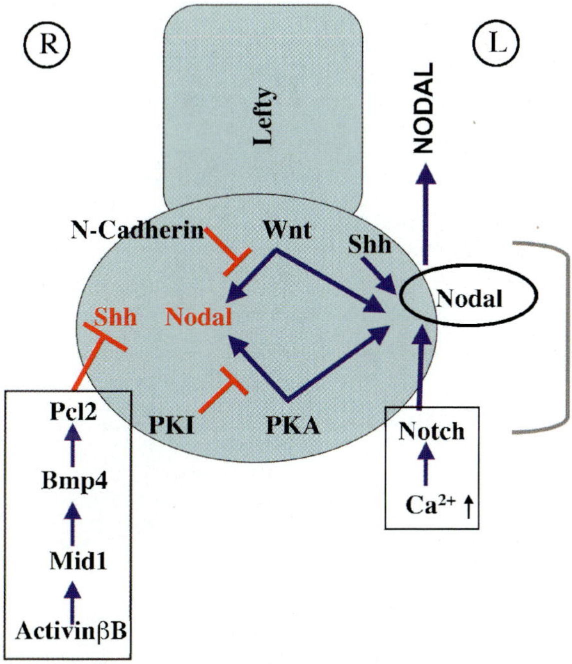

Fig. 7 Schematic diagram depicting the signaling pathways operating in the chick node. Signaling molecules mediating LR asymmetrical patterning in the developing chick embryo. A right side cascade (*left box*) represses *Shh* signaling on the right side of the node (represented by the *colored circle*). Shh on the left induces Nodal on its left-side perinodal domain. Independently, Wnt and PKA activities positively regulate Nodal in the node, but activity on the right side is blocked by N-cadherin and PKI, respectively. Nodal is also activated on the left by a local boost in Notch activity, in turn dependent on a local increase in extracellular calcium. As a consequence of all these regulatory mechanisms, Nodal is only activated on the left side of the node. Negative interactions and blocked expressions are in *red*

Fig. 8A–D Asymmetric gene expressions in the gastrulating chick embryo. **A**, **B**, and **D** are pictures of whole-mount in situ hybridizations showing asymmetric *Shh* (**A**) *Bmp4* (**B**), and *Fgf8* (**D**) expression in Hensen's node, as marked by the *arrows*. **C** Whole-mount in situ hybridization of a stage 6HH embryo showing the left-sided perinodal domain of *Nodal*. Note that *Shh* and *Bmp4* patterns of expression in the node are complementary. All the panels are dorsal views. The *right side* is that of the reader

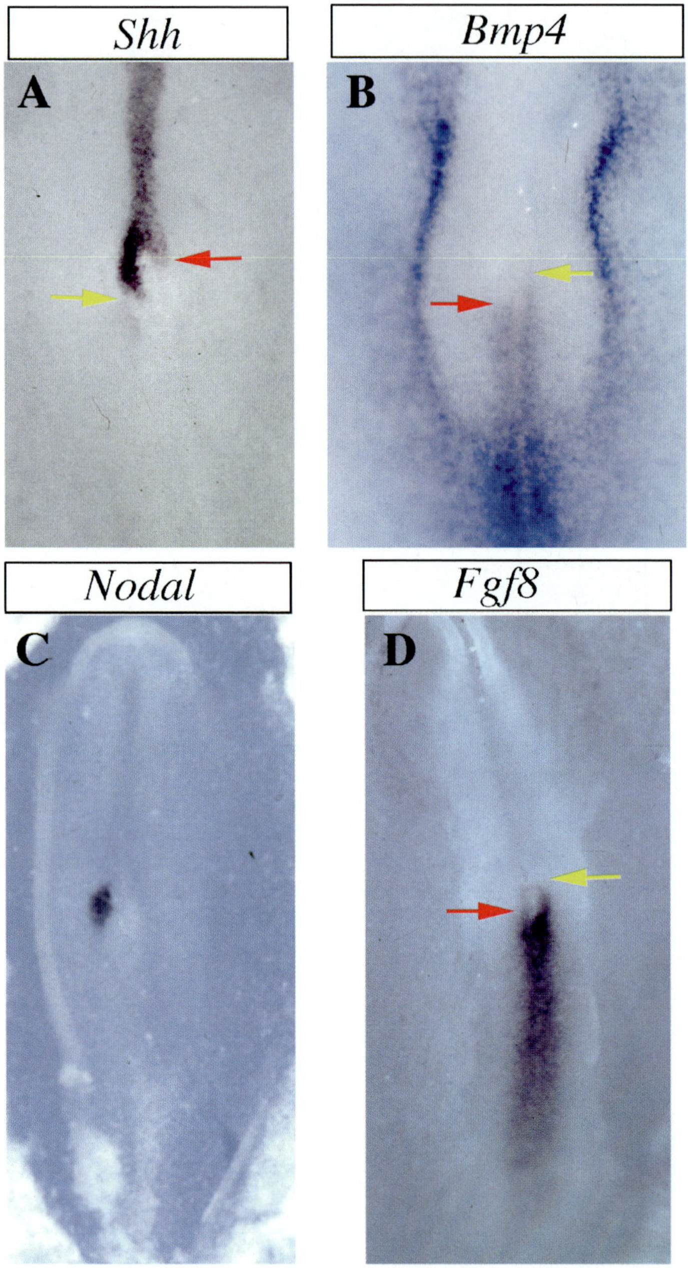
Shh
A
Bmp4
B
Nodal
C
Fgf8
D

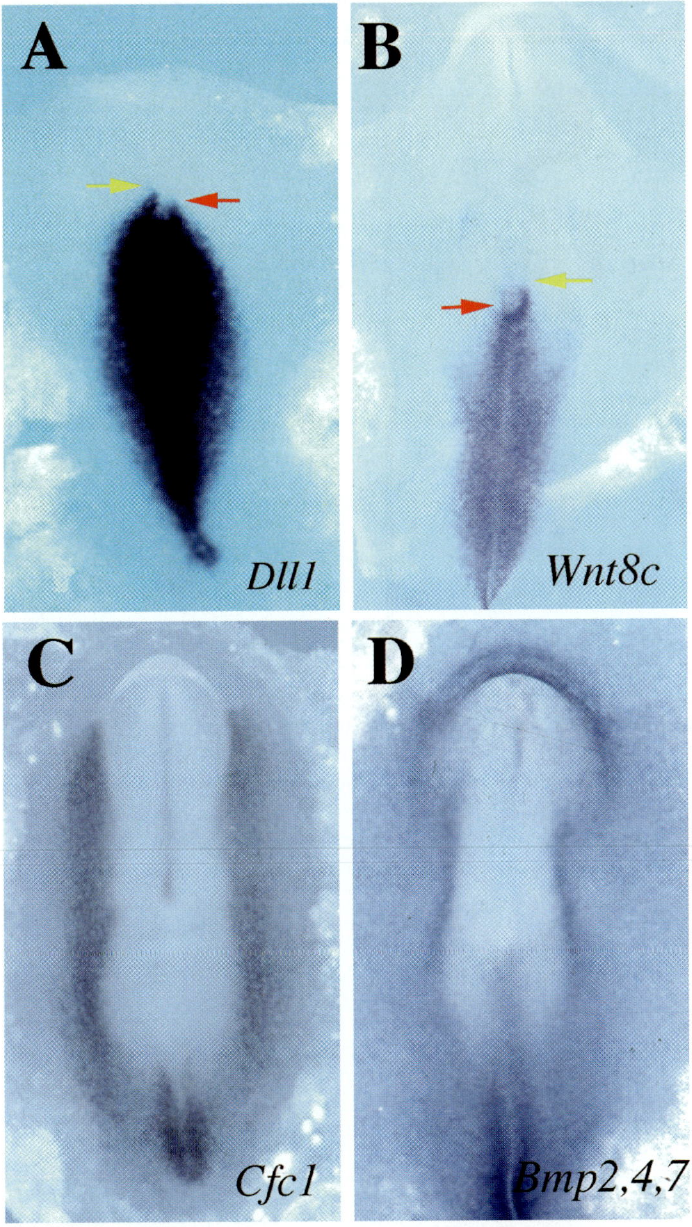

Fig. 9A–D *Dll1*, *Wnt8c*, *Cfc1*, and *Bmp* patterns of expression in the early chick embryo. A and B are pictures of whole-mount in situ hybridizations showing asymmetric *Dll1* (**A**) and *Wnt8c* (**B**) expression in Hensen's node, as indicated by the *arrows*. **C** is a picture of a stage 6HH embryo showing the symmetric pattern of expression of *Cfc1*. **D** is a picture of a stage 6HH embryo hybridized for *Bmp2*, *Bmp4* and *Bmp7* together. The expression of these three genes is also symmetric. All the panels are dorsal views. The *right side* is that of the reader

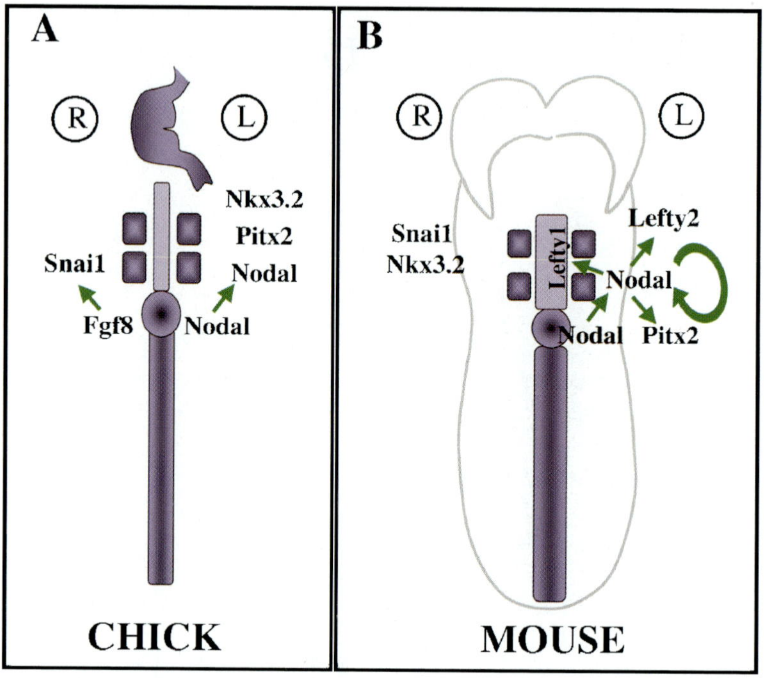

Fig. 10A, B Schematic diagram depicting asymmetric gene expressions in the LPM of chick and mouse embryos. **A** Chick embryo. The asymmetric expression of *Nodal* on the *left side* of the node induces *Nodal, Pitx2* and *Nkx3.2* in the left LPM. On *the right, Fgf8* induces the expression of *Snail1* on the right LPM. **B** Mouse embryo. Nodal spreading from the node, induces *Nodal, Lefty2,* and *Pitx2* expression in the left LPM. Nodal from the LPM induces *Lefty1* in the midline. The mouse right LPM expresses *Snail1* and *Nkx3.2*. Note that *Nkx3.2* is expressed on opposite sides of the LPM in chick and mouse. See text for details

Fig. 11A–D Expression of *Nodal, Pitx2,* and *Snail1* in the chick lateral plate mesoderm. **A** and **B** are whole-mount in situ hybridizations showing the rapid spread of *Nodal* expression on the left LPM from stage 7HH (**A**) to stage 8HH (**A**). At stage 7HH (**A**) *Nodal* expression is just starting whereas a few hours later, at stage 8HH (**B**), it spans along most of the LPM. **C** *Pitx2* is also expressed in the left LPM as shown in this stage 8HH chick embryo, in a pattern similar to *Nodal*. **D** In contrast, *Snail1* is expressed stronger on the right than on the left LPM of a comparable stage 8HH chick embryo. **A** and **B** are ventral views, **C** and **D** are dorsal views. Left (*L*) and right (*R*) as indicated

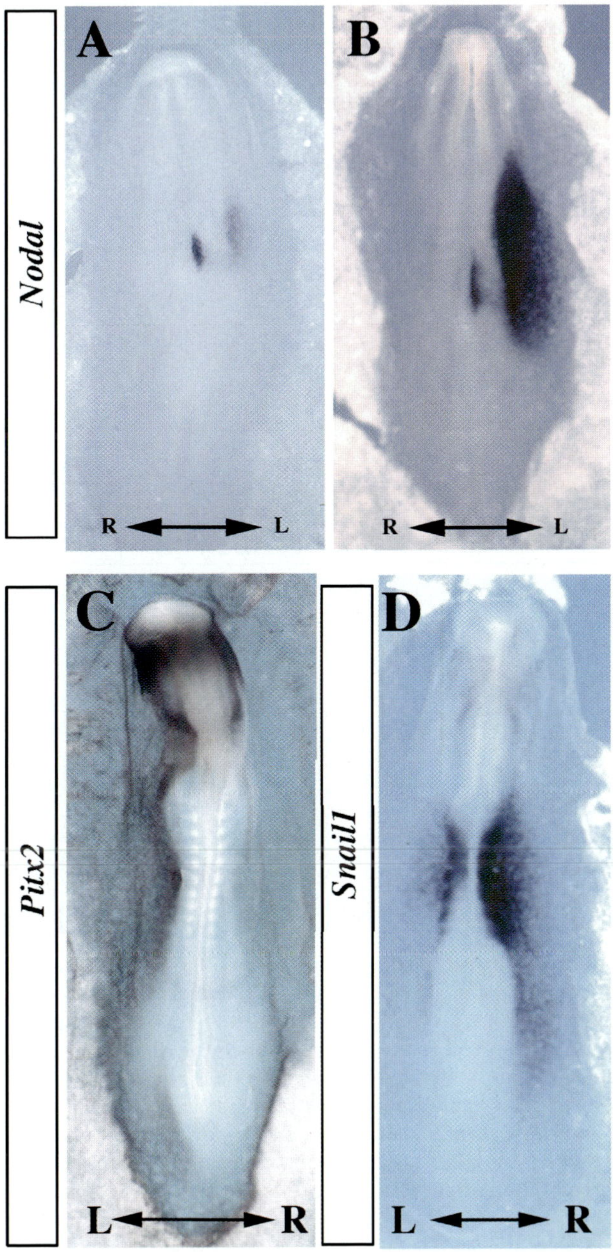
Nodal
A
R L
B
R L
Pitx2
C
L R
Snail1
D
L R

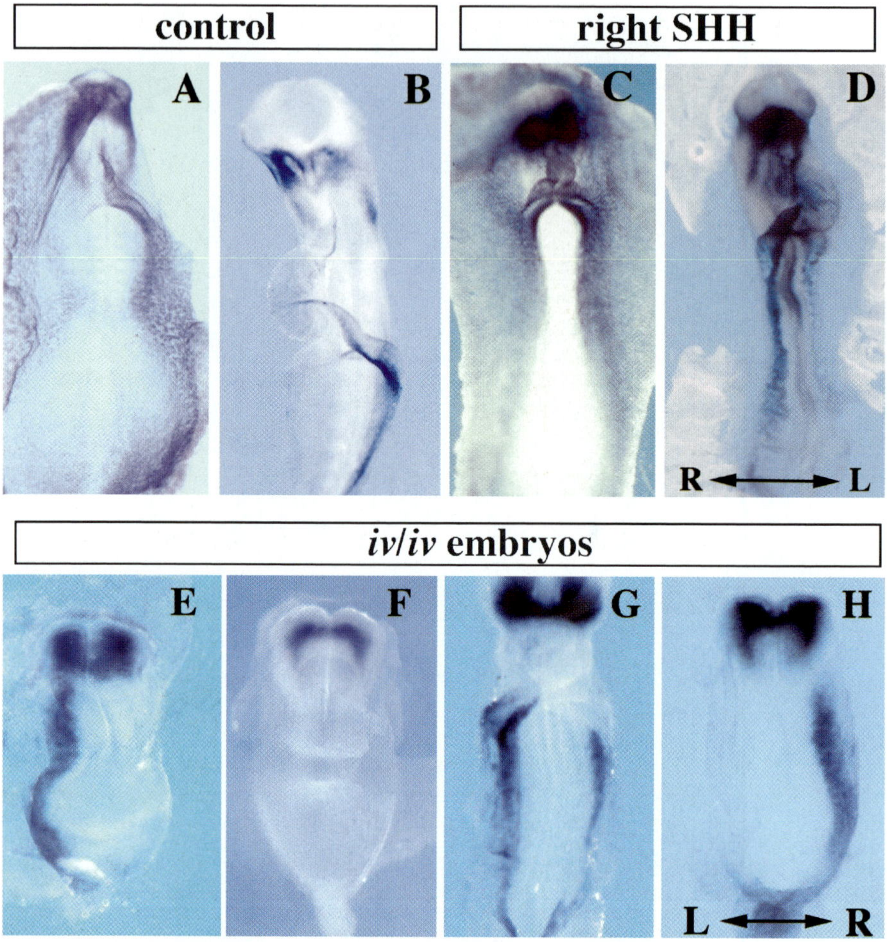

Fig. 12A–H *Pitx2* expression in the developing heart; after SHH ectopic application and in *iv/iv* mouse embryos. **A** Stage 9HH embryo showing normal *Pitx2* expression on the left side of the tubular heart and on the left side of the splanchnopleure. **B** During heart looping, the left-sided expression of *Pitx2* becomes anterior due to the rotation of the heart. **C** Bilateral *Pitx2* expression in an embryo treated with SHH on the right side of Hensen's node at stage 5HH. **D** Left-looped heart caused by the ectopic application of SHH on the right side of Hensen's node and subsequent bilateral *Pitx2* expression. All the panels are ventral views. **E–H** Pattern of expression of *Pitx2* in *iv/iv* embryos. In wild-type embryos it can be left (**E**), absent (**F**), bilateral (**G**), or right (**H**). Bilateral expression may not be strictly symmetric (**G**). Panels **A–D** are ventral views with the LR axis indicated in (**A**). Panels **E–H** show dorsal views with the LR axis indicated in **E**

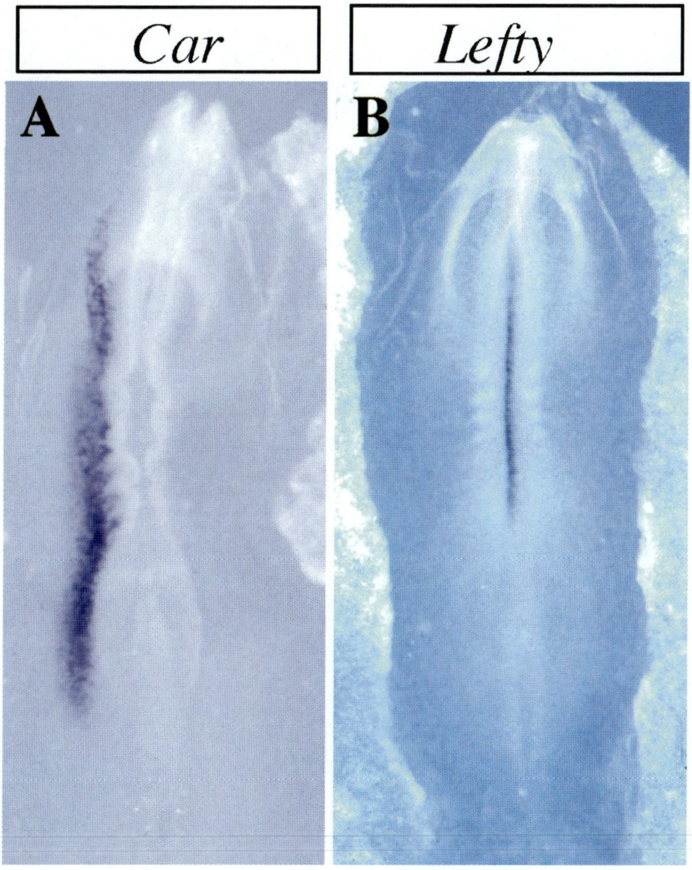

Fig. 13A, B Expression of *Car* and *Lefty1* in early chick embryos. **A** Stage 8HH embryo showing *Car* expression on the left LPM in a pattern similar, but not identical, to that of Nodal. **B** Midline expression of *Lefty1* in a stage 8HH embryo

Fig. 14A–D *Pitx2* and *Nodal* expression in spontaneously occurring conjoined twins. **A** Spontaneously occurring chick conjoined embryos with oblique AP axes hybridized for *Pitx2*. The left-side twin express *Pitx2* at normal level in the left LPM and at low level in the right LPM (*arrow*). The right-side twin express Pitx2 at low level but bilaterally (*arrows*). **B** Spontaneously occurring chick conjoined embryos with almost parallel primitive streaks hybridized for *Nodal*. The left-side embryo expresses *Nodal* normally on the left. Unexpectedly, a weak expression of *Nodal* is also detected in the left LPM of the right twin. **C** Cartoon depicting the presumed origin of the embryo in **A**. In oblique primitive streaks the left-side signaling cascade (*L*) of the right embryo causes bilateral expression of *Pitx2* in the left embryo. **D** Cartoon depicting the presumed origin of the embryo in (**B**). In parallel primitive streaks the right-side cascade (*R*) of the left embryo inhibits the expression of Nodal on the left side of the right embryo. The LR axis is indicated in the *panels*

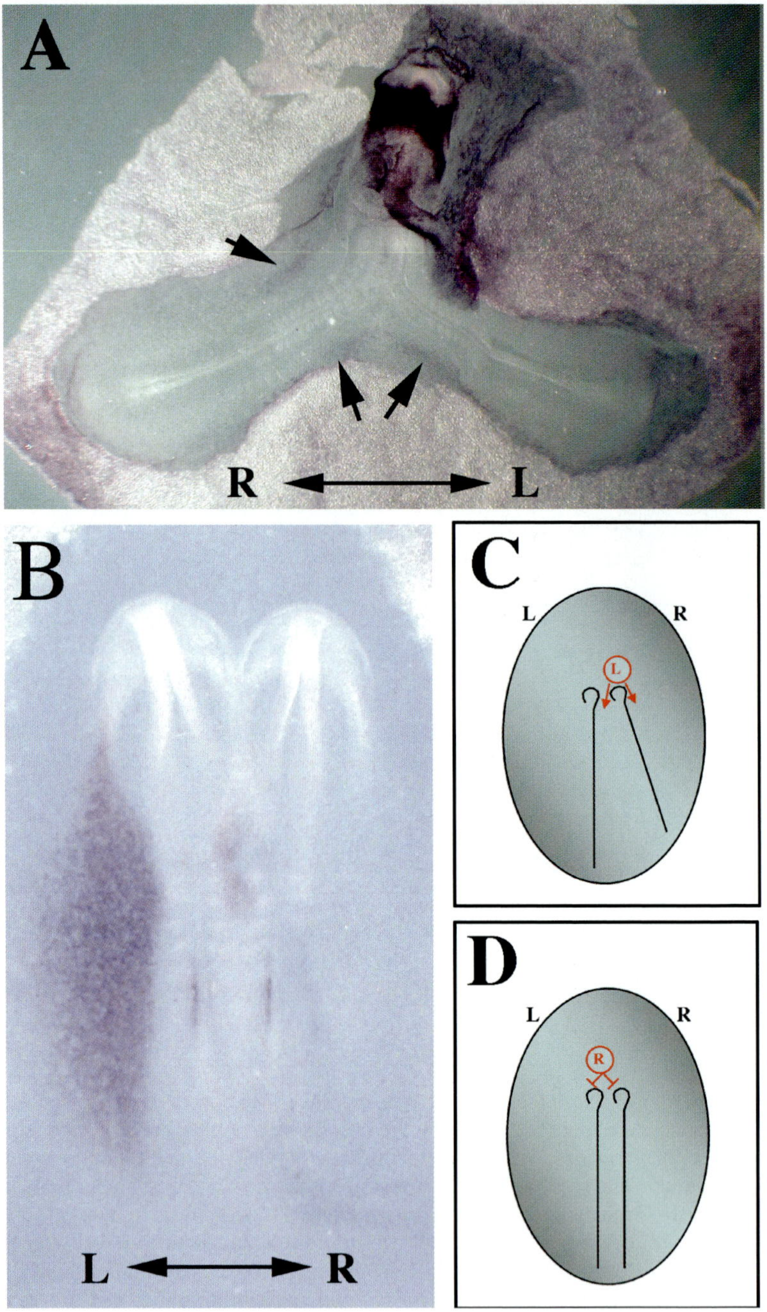
A
R
L
B
L
R
C
L
R
L
D
L
R
R

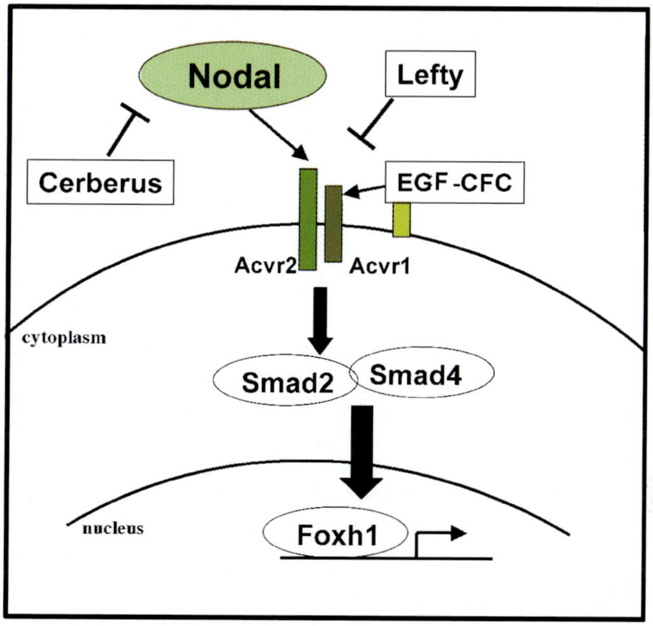

Fig. 15 Schematic diagram of Nodal signaling from cell membrane to the nucleus. The ligand Nodal and the antagonists Cerberus and Lefty are depicted outside the cell membrane. The receptor type I (Acvr1), the receptor type II (Acvr2) and the co-receptor EGF-CFC are depicted in the cell membrane. In response to ligand binding, phosphorylated Smad2 associated with the co-Smad (Smad4) translocates to the nucleus, and activates downstream target genes in association with the transcription factor Foxh1. The *arrows* indicate signal flow

References

Adachi H, Saijoh Y, Mochida K, Ohishi S, Hashiguchi H, Hirao A and Hamada H (1999) Determination of left/right asymmetric expression of nodal by a left side-specific enhancer with sequence similarity to a lefty-2 enhancer Genes Dev 13, 1589–600

Adamson ED, Minchiotti G and Salomon DS (2002) Cripto: a tumor growth factor and more J Cell Physiol 190, 267–78

Afzelius BA (1976) A human syndrome caused by immotile cilia Science 193, 317–9

Afzelius BA (1979) The immotile-cilia syndrome and other ciliary diseases Int Rev Exp Pathol 19, 1–43

Afzelius BA (1985) The immotile-cilia syndrome: a microtubule-associated defect CRC Crit Rev Biochem 19, 63–87

Aird I (1959) Conjoined twins Br. Med. J. 1, 1313–1325

Akiyama H, Shigeno C, Hiraki Y, Shukunami C, Kohno H, Akagi M, Konishi J and Nakamura T (1997) Cloning of a mouse smoothened cDNA and expression patterns of hedgehog signalling molecules during chondrogenesis and cartilage differentiation in clonal mouse EC cells, ATDC5 Biochem Biophys Res Commun 235, 142–7

Amendt BA, Semina EV and Alward WL (2000) Rieger syndrome: a clinical, molecular, and biochemical analysis Cell Mol Life Sci 57, 1652–66

Ang SL and Rossant J(1994) HNF-3 beta is essential for node and notochord formation in mouse development Cell 78, 561–74

Ang SL, Wierda A, Wong D, Stevens KA, Cascio S, Rossant J and Zaret KS (1993) The formation and maintenance of the definitive endoderm lineage in the mouse: involvement of HNF3/forkhead proteins Development 119, 1301–15

Antin PB, Taylor RG and Yatskievych T (1994) Precardiac mesoderm is specified during gastrulation in quail Dev Dyn 200, 144–54

Aplan PD, Lombardi DP, Ginsberg AM, Cossman J, Bertness VL and Kirsch IR (1990) Disruption of the human SCL locus by "illegitimate" V-(D)-J recombinase activity Science 250, 1426–9

Bachiller D, Klingensmith J, Kemp C, Belo JA, Anderson RM, May SR, McMahon JA, McMahon AP, Harland RM, Rossant J et al. (2000) The organizer factors Chordin and Noggin are required for mouse forebrain development Nature 403, 658–61

Bachiller D, Klingensmith J, Shneyder N, Tran U, Anderson R, Rossant J and De Robertis EM (2003) The role of chordin/Bmp signals in mammalian pharyngeal development and DiGeorge syndrome Development 130, 3567–78

Bamford RN, Roessler E, Burdine RD, Saplakoglu U, dela Cruz J, Splitt M, Goodship JA, Towbin J, Bowers P, Ferrero GB et al. (2000) Loss-of-function mutations in the EGF-CFC gene CFC1 are associated with human left–right laterality defects Nat Genet 26, 365–9

Bamforth SD, Braganca J, Eloranta JJ, Murdoch JN, Marques FI, Kranc KR, Farza H, Henderson DJ, Hurst HC and Bhattacharya S (2001) Cardiac malformations, adrenal agenesis, neural crest defects and exencephaly in mice lacking Cited2, a new Tfap2 co-activator Nat Genet 29, 469–74

Bamforth SD, Braganca J, Farthing CR, Schneider JE, Broadbent C, Michell AC, Clarke K, Neubauer S, Norris D, Brown NA et al. (2004) Cited2 controls left–right patterning and heart development through a Nodal-Pitx2c pathway Nat Genet 36, 1189–96

Barrallo-Gimeno A and Nieto MA (2005) The Snail genes as inducers of cell movement and survival: implications in development and cancer Development 132, 3151–61

Bartoloni L, Blouin JL, Maiti AK, Sainsbury A, Rossier C, Gehrig C, She JX, Marron MP, Lander ES, Meeks M et al. (2001) Axonemal beta heavy chain dynein DNAH9: cDNA sequence, genomic structure, and investigation of its role in primary ciliary dyskinesia Genomics 72, 21–33

Bartoloni L, Blouin JL, Pan Y, Gehrig C, Maiti AK, Scamuffa N, Rossier C, Jorissen M, Armengot M, Meeks M et al. (2002) Mutations in the DNAH11 (axonemal heavy chain dynein type 11) gene cause one form of situs inversus totalis and most likely primary ciliary dyskinesia Proc Natl Acad Sci U S A 99, 10282–6

Beck S, Le Good JA, Guzman M, Ben Haim N, Roy K, Beermann F and Constam DB (2002) Extraembryonic proteases regulate Nodal signalling during gastrulation Nat Cell Biol 4, 981–5

Beddington RS (1994) Induction of a second neural axis by the mouse node Development 120, 613–20

Beddington RS and Robertson EJ (1999) Axis development and early asymmetry in mammals Cell 96, 195–209

Bellomo D, Lander A, Harragan I and Brown NA (1996) Cell proliferation in mammalian gastrulation: the ventral node and notochord are relatively quiescent Dev Dyn 205, 471–85

Belloni E, Muenke M, Roessler E, Traverso G, Siegel-Bartelt J, Frumkin A, Mitchell HF, Donis-Keller H, Helms C, Hing AV et al. (1996) Identification of Sonic hedgehog as a candidate gene responsible for holoprosencephaly Nat Genet 14, 353–6

Berg C, Geipel A, Smrcek J, Krapp M, Germer U, Kohl T, Gembruch U and Baschat AA (2003) Prenatal diagnosis of cardiosplenic syndromes: a 10-year experience Ultrasound Obstet Gynecol 22, 451–9

Bisgrove BW, Essner JJ and Yost HJ (1999) Regulation of midline development by antagonism of lefty and nodal signaling Development 126, 3253–62

Boettger T, Wittler L and Kessel M (1999) FGF8 functions in the specification of the right body side of the chick Curr Biol 9, 277–80

Bowers PN, Brueckner M and Yost HJ (1996) The genetics of left–right development and heterotaxia Semin Perinatol 20, 577–88

Braganca J, Eloranta JJ, Bamforth SD, Ibbitt JC, Hurst HC and Bhattacharya S (2003) Physical and functional interactions among AP-2 transcription factors, p300/CREB-binding protein, and CITED2 J Biol Chem 278, 16021–9

Branford WW, Essner JJ and Yost HJ (2000) Regulation of gut and heart left–right asymmetry by context-dependent interactions between xenopus lefty and BMP4 signaling Dev Biol 223, 291–306

Branford WW and Yost HJ (2002) Lefty-dependent inhibition of Nodal- and Wnt-responsive organizer gene expression is essential for normal gastrulation Curr Biol 12, 2136–41

Brennan J, Norris DP and Robertson EJ (2002) Nodal activity in the node governs left–right asymmetry Genes Dev 16, 2339–44

Britz-Cunningham SH, Shah MM, Zuppan CW and Fletcher WH (1995) Mutations of the Connexin43 gap-junction gene in patients with heart malformations and defects of laterality N Engl J Med 332, 1323–9

Brown NA and Wolpert L (1990) The development of handedness in left/right asymmetry Development 109, 1–9

Brueckner M (2001) Cilia propel the embryo in the right direction Am J Med Genet 101, 339–44

Brueckner M, D'Eustachio P and Horwich AL (1989) Linkage mapping of a mouse gene, iv, that controls left–right asymmetry of the heart and viscera Proc Natl Acad Sci U S A 86, 5035–8

Bunney TD, De Boer AH and Levin M (2003) Fusicoccin signaling reveals 14-3-3 protein function as a novel step in left–right patterning during amphibian embryogenesis Development 130, 4847–58

Burdine RD and Schier AF (2000) Conserved and divergent mechanisms in left–right axis formation Genes Dev 14, 763–76

Burn J (1991) Disturbance of morphological laterality in humans Ciba Found Symp 162, 282–96; discussion 296–9

Campione M, Ros MA, Icardo JM, Piedra E, Christoffels VM, Schweickert A, Blum M, Franco D and Moorman AF (2001) Pitx2 expression defines a left cardiac lineage of cells: evidence for atrial and ventricular molecular isomerism in the iv/iv mice Dev Biol 231, 252–64

Campione M, Steinbeisser H, Schweickert A, Deissler K, van Bebber F, Lowe LA, Nowotschin S, Viebahn C, Haffter P, Kuehn MR et al. (1999) The homeobox gene Pitx2: mediator of asymmetric left–right signaling in vertebrate heart and gut looping Development 126, 1225–34

Capdevila J, Vogan KJ, Tabin CJ and Izpisua Belmonte JC (2000) Mechanisms of left–right determination in vertebrates Cell 101, 9–21

Carlson BM (2000) Human Embryology and Developmental Biology. London: Harcourt

Chang C, Wilson PA, Mathews LS and Hemmati-Brivanlou A (1997) A Xenopus type I activin receptor mediates mesodermal but not neural specification during embryogenesis Development 124, 827–37

Chang H, Zwijsen A, Vogel H, Huylebroeck D and Matzuk MM (2000) Smad5 is essential for left–right asymmetry in mice Dev Biol 219, 71–8

Chazaud C, Chambon P and Dolle P (1999) Retinoic acid is required in the mouse embryo for left–right asymmetry determination and heart morphogenesis Development 126, 2589–96

Chen JN, van Eeden FJ, Warren KS, Chin A, Nusslein-Volhard C, Haffter P and Fishman MC (1997) Left–right pattern of cardiac BMP4 may drive asymmetry of the heart in zebrafish Development 124, 4373–82

Chen Y and Schier AF (2002) Lefty proteins are long-range inhibitors of squint-mediated nodal signaling Curr Biol 12, 2124–8

Cheng SK, Olale F, Bennett JT, Brivanlou AH and Schier AF (2003) EGF-CFC proteins are essential coreceptors for the TGF-beta signals Vg1 and GDF1 Genes Dev 17, 31–6

Cheng SK, Olale F, Brivanlou AH and Schier AF (2004) Lefty blocks a subset of TGFbeta signals by antagonizing EGF-CFC coreceptors PLoS Biol 2 E30

Chiang C, Litingtung Y, Lee E, Young KE, Corden JL, Westphal H and Beachy PA (1996) Cyclopia and defective axial patterning in mice lacking Sonic hedgehog gene function Nature 383, 407–13

Chin AJ, Tsang M and Weinberg ES (2000) Heart and gut chiralities are controlled independently from initial heart position in the developing zebrafish Dev Biol 227, 403–21

Ciccodicola A, Dono R, Obici S, Simeone A, Zollo M and Persico MG (1989) Molecular characterization of a gene of the 'EGF family' expressed in undifferentiated human NTERA2 teratocarcinoma cells Embo J 8, 1987–91

Colas JF and Schoenwolf GC (2000) Subtractive hybridization identifies chick-cripto, a novel EGF-CFC ortholog expressed during gastrulation, neurulation and early cardiogenesis Gene 255, 205–17

Collignon J, Varlet I and Robertson EJ (1996) Relationship between asymmetric nodal expression and the direction of embryonic turning Nature 381, 155–8

Connolly DJ, Patel K and Cooke J (1997) Chick noggin is expressed in the organizer and neural plate during axial development, but offers no evidence of involvement in primary axis formation Int J Dev Biol 41, 389–96

Constam DB and Robertson EJ (2000a) SPC4/PACE4 regulates a TGFbeta signaling network during axis formation Genes Dev 14, 1146–55

Constam DB and Robertson EJ (2000b) Tissue-specific requirements for the proprotein convertase furin/SPC1 during embryonic turning and heart looping Development 127, 245–54

Cooke J (1995) Vertebrate embryo handedness Nature 374, 681

Cooke J (2004) The evolutionary origins and significance of vertebrate left–right organisation Bioessays 26, 413–21

Cooper AF, Yu KP, Brueckner M, Brailey LL, Johnson L, McGrath JM and Bale AE (2005) Cardiac and CNS defects in a mouse with targeted disruption of suppressor of fused Development 132, 4407–17

Corbit KC, Aanstad P, Singla V, Norman AR, Stainier DY and Reiter JF (2005) Vertebrate Smoothened functions at the primary cilium Nature 437, 1018–21

Cox CJ, Espinoza HM, McWilliams B, Chappell K, Morton L, Hjalt TA, Semina EV and Amendt BA (2002) Differential regulation of gene expression by PITX2 isoforms J Biol Chem 277, 25001–10

Danos MC and Yost HJ (1995) Linkage of cardiac left–right asymmetry and dorsal-anterior development in Xenopus Development 121, 1467–74

Danos MC and Yost HJ (1996) Role of notochord in specification of cardiac left–right orientation in zebrafish and Xenopus Dev Biol 177, 96–103

Dasgupta C, Martinez AM, Zuppan CW, Shah MM, Bailey LL and Fletcher WH (2001) Identification of connexin43 (alpha1) gap junction gene mutations in patients with hypoplastic left heart syndrome by denaturing gradient gel electrophoresis (DGGE) Mutat Res 479, 173–86

Dathe V, Gamel A, Manner J, Brand-Saberi B and Christ B (2002) Morphological left–right asymmetry of Hensen's node precedes the asymmetric expression of Shh and Fgf8 in the chick embryo Anat Embryol (Berl) 205, 343–54

Davidson BP, Kinder SJ, Steiner K, Schoenwolf GC and Tam PP (1999) Impact of node ablation on the morphogenesis of the body axis and the lateral asymmetry of the mouse embryo during early organogenesis Dev Biol 211, 11–26

Davies M and Guest PJ (2003) Developmental abnormalities of the great vessels of the thorax and their embryological basis Br J Radiol 76, 491–502

De Marco V, Burkhard P, Le Bot N, Vernos I and Hoenger A (2001) Analysis of heterodimer formation by Xklp3A/B, a newly cloned kinesin-II from Xenopus laevis Embo J 20, 3370–9

Debrus S, Tuffery S, Matsuoka R, Galal O, Sarda P, Sauer U, Bozio A, Tanman B, Toutain A, Claustres M et al. (1997) Lack of evidence for connexin 43 gene mutations in human autosomal recessive lateralization defects J Mol Cell Cardiol 29, 1423–31

Dersch H and Zile MH (1993) Induction of normal cardiovascular development in the vitamin A-deprived quail embryo by natural retinoids Dev Biol 160, 424–33
Dickman ED and Smith SM (1996) Selective regulation of cardiomyocyte gene expression and cardiac morphogenesis by retinoic acid Dev Dyn 206, 39–48
Dono R, Montuori N, Rocchi M, De Ponti-Zilli L, Ciccodicola A and Persico MG (1991) Isolation and characterization of the CRIPTO autosomal gene and its X-linked related sequence Am J Hum Genet 49, 555–65
Dono R, Scalera L, Pacifico F, Acampora D, Persico MG and Simeone A (1993) The murine cripto gene: expression during mesoderm induction and early heart morphogenesis Development 118, 1157–68
Duboc V, Rottinger E, Lapraz F, Besnardeau L and Lepage T (2005) Left–right asymmetry in the sea urchin embryo is regulated by nodal signaling on the right side Dev Cell 9, 147–58
Dufort D, Schwartz L, Harpal K and Rossant J (1998) The transcription factor HNF3beta is required in visceral endoderm for normal primitive streak morphogenesis Development 125, 3015–25
Dunwoodie SL, Rodriguez TA and Beddington RS (1998) Msg1 and Mrg1, founding members of a gene family, show distinct patterns of gene expression during mouse embryogenesis Mech Dev 72, 27–40
Echelard Y, Epstein DJ, St-Jacques B, Shen L, Mohler J, McMahon JA and McMahon AP (1993) Sonic hedgehog, a member of a family of putative signaling molecules, is implicated in the regulation of CNS polarity Cell 75, 1417–30
Eley L, Turnpenny L, Yates LM, Craighead AS, Morgan D, Whistler C, Goodship JA and Strachan T (2004) A perspective on inversin Cell Biol Int 28, 119–24
Ericson J, Muhr J, Jessell TM and Edlund T (1995) Sonic hedgehog: a common signal for ventral patterning along the rostrocaudal axis of the neural tube Int J Dev Biol 39, 809–16
Essner JJ, Amack JD, Nyholm MK, Harris EB and Yost HJ (2005) Kupffer's vesicle is a ciliated organ of asymmetry in the zebrafish embryo that initiates left–right development of the brain, heart and gut Development 132, 1247–60
Essner JJ, Vogan KJ, Wagner MK, Tabin CJ, Yost HJ and Brueckner M (2002) Conserved function for embryonic nodal cilia Nature 418, 37–8
Evans WH and Martin PE (2002) Gap junctions: structure and function (Review) Mol Membr Biol 19, 121–36
Faisst AM, Alvarez-Bolado G, Treichel D and Gruss P (2002) Rotatin is a novel gene required for axial rotation and left–right specification in mouse embryos Mech Dev 113, 15–28
Ferencz C, Rubin JD, McCarter RJ, Brenner JI, Neill CA, Perry LW, Hepner SI and Downing JW (1985) Congenital heart disease: prevalence at live birth. The Baltimore-Washington Infant Study Am J Epidemiol 121, 31–6
Fischer A, Viebahn C and Blum M (2002) FGF8 acts as a right determinant during establishment of the left–right axis in the rabbit Curr Biol 12, 1807–16
Franco D and Campione M (2003) The role of Pitx2 during cardiac development. Linking left–right signaling and congenital heart diseases Trends Cardiovasc Med 13, 157–63
Franco D and Icardo JM (2001) Molecular characterization of the ventricular conduction system in the developing mouse heart: topographical correlation in normal and congenitally malformed hearts Cardiovasc Res 49, 417–29
Fujinaga M (1997) Development of sidedness of asymmetric body structures in vertebrates Int J Dev Biol 41, 153–86

Fujiwara T, Dehart DB, Sulik KK and Hogan BL (2002) Distinct requirements for extra-embryonic and embryonic bone morphogenetic protein 4 in the formation of the node and primitive streak and coordination of left–right asymmetry in the mouse Development 129, 4685–96

Fukumoto T, Kema IP and Levin M (2005) Serotonin signaling is a very early step in patterning of the left–right axis in chick and frog embryos Curr Biol 15, 794–803

Gage PJ, Suh H and Camper SA (1999) Dosage requirement of Pitx2 for development of multiple organs Development 126, 4643–51

Gaio U, Schweickert A, Fischer A, Garratt AN, Muller T, Ozcelik C, Lankes W, Strehle M, Britsch S, Blum M et al. (1999) A role of the cryptic gene in the correct establishment of the left–right axis Curr Biol 9, 1339–42

Garcia-Castro MI, Vielmetter E and Bronner-Fraser M (2000) N-Cadherin, a cell adhesion molecule involved in establishment of embryonic left–right asymmetry Science 288, 1047–51

Gebbia M, Ferrero GB, Pilia G, Bassi MT, Aylsworth A, Penman-Splitt M, Bird LM, Bamforth JS, Burn J, Schlessinger D et al. (1997) X-linked situs abnormalities result from mutations in ZIC3 Nat Genet 17, 305–8

Gebbia M, Towbin JA and Casey B (1996) Failure to detect connexin43 mutations in 38 cases of sporadic and familial heterotaxy Circulation 94, 1909–12

Goldmuntz E, Bamford R, Karkera JD, dela Cruz J, Roessler E and Muenke M (2002) CFC1 mutations in patients with transposition of the great arteries and double-outlet right ventricle Am J Hum Genet 70, 776–80

Goldstein AM, Ticho BS and Fishman MC (1998) Patterning the heart's left–right axis: from zebrafish to man Dev Genet 22, 278–87

Granata A and Quaderi NA (2003) The Opitz syndrome gene MID1 is essential for establishing asymmetric gene expression in Hensen's node Dev Biol 258, 397–405

Granata A and Quaderi NA (2005) Asymmetric expression of Gli transcription factors in Hensen's node Gene Expr Patterns 5, 529–31

Granata A, Savery D, Hazan J, Cheung BM, Lumsden A and Quaderi NA (2005) Evidence of functional redundancy between MID proteins: implications for the presentation of Opitz syndrome Dev Biol 277, 417–24

Gritsman K, Zhang J, Cheng S, Heckscher E, Talbot WS and Schier AF (1999) The EGF-CFC protein one-eyed pinhead is essential for nodal signaling Cell 97, 121–32

Guichard C, Harricane MC, Lafitte JJ, Godard P, Zaegel M, Tack V, Lalau G and Bouvagnet P (2001) Axonemal dynein intermediate-chain gene (DNAI1) mutations result in situs inversus and primary ciliary dyskinesia (Kartagener syndrome) Am J Hum Genet 68, 1030–5

Gurdon JB (2005) Sinistral snails and gentlemen scientists Cell 123, 751–3

Guthrie S, Turin L and Warner A (1988) Patterns of junctional communication during development of the early amphibian embryo Development 103, 769–83

Hamada H, Meno C, Watanabe D and Saijoh Y (2002) Establishment of vertebrate left–right asymmetry Nat Rev Genet 3, 103–13

Hamburger V and Hamilton HL (1951) A series of normal stages in the development of the chick embryo Journal of Morphology 88, 49–92

Hara K (1978) Spermann's organizer in birds. In Organizer: a Milestone of a Half-Century from, vol. Hara K (1978) Spermann's organizer in birds. In Organizer: a Milestone of a Half-Century from, (ed. O Nakamura and S Toivonen), pp. 221–265. Amsterdam: Elsevier/NorthHolland

Haycraft CJ, Banizs B, Aydin-Son Y, Zhang Q, Michaud EJ and Yoder BK (2005) Gli2 and Gli3 localize to cilia and require the intraflagellar transport protein polaris for processing and function PLoS Genet 1, e53

Hecksher-Sorensen J, Watson RP, Lettice LA, Serup P, Eley L, De Angelis C, Ahlgren U and Hill RE (2004) The splanchnic mesodermal plate directs spleen and pancreatic laterality, and is regulated by Bapx1/Nkx3.2 Development 131, 4665–75

Heymer J, Kuehn M and Ruther U (1997) The expression pattern of nodal and lefty in the mouse mutant Ft suggests a function in the establishment of handedness Mech Dev 66, 5–11

Hirokawa N, Tanaka Y, Okada Y and Takeda S (2006) Nodal flow and the generation of left–right asymmetry Cell 125, 33–45

Hogan BL (1996) Bone morphogenetic proteins in development Curr Opin Genet Dev 6, 432–8

Hoodless PA, Pye M, Chazaud C, Labbe E, Attisano L, Rossant J and Wrana JL (2001) FoxH1 (Fast) functions to specify the anterior primitive streak in the mouse Genes Dev 15, 1257–71

Hooper JE and Scott MP (2005) Communicating with Hedgehogs Nat Rev Mol Cell Biol 6, 306–17

Horvath J, Fliegauf M, Olbrich H, Kispert A, King SM, Mitchison H, Zariwala MA, Knowles MR, Sudbrak R, Fekete G et al. (2005) Identification and analysis of axonemal dynein light chain 1 in primary ciliary dyskinesia patients Am J Respir Cell Mol Biol 33, 41–7

Hou J, Yashiro K, Okazaki Y, Saijoh Y, Hayashizaki Y and Hamada H (2004) Identification of a novel left–right asymmetrically expressed gene in the mouse belonging to the BPI/PLUNC superfamily Dev Dyn 229, 373–9

Huangfu D and Anderson KV (2005) Cilia and Hedgehog responsiveness in the mouse Proc Natl Acad Sci U S A 102, 11325–30

Huangfu D, Liu A, Rakeman AS, Murcia NS, Niswander L and Anderson KV (2003) Hedgehog signalling in the mouse requires intraflagellar transport proteins Nature 426, 83–7

Hummel KP and Chapman DB (1959) Visceral inversion and associated anomalies in the mouse J. Hered 50, 9–13

Hutson MR and Kirby ML (2003) Neural crest and cardiovascular development: a 20-year perspective Birth Defects Res C Embryo Today 69, 2–13

Hyatt BA, Lohr JL and Yost HJ (1996) Initiation of vertebrate left–right axis formation by maternal Vg1 Nature 384, 62–5

Hyatt BA and Yost HJ (1998) The left–right coordinator: the role of Vg1 in organizing left–right axis formation Cell 93, 37–46

Ibanez-Tallon I, Gorokhova S and Heintz N (2002) Loss of function of axonemal dynein Mdnah5 causes primary ciliary dyskinesia and hydrocephalus Hum Mol Genet 11, 715–21

Icardo JM (1996) Developmental biology of the vertebrate heart J Exp Zool 275, 144–61

Icardo JM (1997) Morphogenesis of vertebrate hearts. In Development of Cardiovascular Systems. Molecules to Organisms, vol. Icardo JM (1997) Morphogenesis of vertebrate hearts. In Development of Cardiovascular Systems. Molecules to Organisms, (ed. W Burggren and B Keller), pp. 114–126. New York: Cambridge University Press

Icardo JM, Guerrero A, Duran AC, Domezain A, Colvee E and Sans-Coma V (2004) The development of the sturgeon heart Anat Embryol (Berl) 208, 439–49

Icardo JM and Ojeda JL (1984) Effects of colchicine on the formation and looping of the tubular heart of the embryonic chick Acta Anat (Basel) 119, 1–9

Icardo JM and Sanchez de Vega MJ (1991) Spectrum of heart malformations in mice with situs solitus, situs inversus, and associated visceral heterotaxy Circulation 84, 2547–58

Igarashi P and Somlo S (2002) Genetics and pathogenesis of polycystic kidney disease J Am Soc Nephrol 13, 2384–98

Ingham PW and McMahon AP (2001) Hedgehog signaling in animal development: paradigms and principles Genes Dev 15, 3059–87

Isaac A, Sargent MG and Cooke J (1997) Control of vertebrate left–right asymmetry by a snail-related zinc finger gene Science 275, 1301–4

Ishimaru Y, Yoshioka H, Tao H, Thisse B, Thisse C, C VEW, Hamada H, Ohuchi H and Noji S (2000) Asymmetric expression of antivin/lefty1 in the early chick embryo Mech Dev 90, 115–8

Izraeli S, Colaizzo-Anas T, Bertness VL, Mani K, Aplan PD and Kirsch IR (1997) Expression of the SIL gene is correlated with growth induction and cellular proliferation Cell Growth Differ 8, 1171–9

Izraeli S, Lowe LA, Bertness VL, Campaner S, Hahn H, Kirsch IR and Kuehn MR (2001) Genetic evidence that Sil is required for the Sonic Hedgehog response pathway Genesis 31, 72–7

Izraeli S, Lowe LA, Bertness VL, Good DJ, Dorward DW, Kirsch IR and Kuehn MR (1999) The SIL gene is required for mouse embryonic axial development and left–right specification Nature 399, 691–4

Kalderon D (2000) Transducing the hedgehog signal Cell 103, 371–4

Kaufman MH (2004) The embryology of conjoined twins Childs Nerv Syst 20, 508–25

Kawakami M and Nakanishi N (2001) The role of an endogenous PKA inhibitor PKIalpha, in organizing left–right axis formation Development 128, 2509–15

Kawakami Y, Raya A, Raya RM, Rodriguez-Esteban C and Belmonte JC (2005) Retinoic acid signalling links left–right asymmetric patterning and bilaterally symmetric somitogenesis in the zebrafish embryo Nature 435, 165–71

Kelly KA, Wei Y and Mikawa T (2002) Cell death along the embryo midline regulates left–right sidedness Dev Dyn 224, 238–44

Kelly RG, Brown NA and Buckingham ME (2001) The arterial pole of the mouse heart forms from Fgf10-expressing cells in pharyngeal mesoderm Dev Cell 1, 435–40

King T, Beddington RS and Brown NA (1998) The role of the brachyury gene in heart development and left–right specification in the mouse Mech Dev 79, 29–37

Kingsley DM (1994) The TGF-beta superfamily: new members, new receptors, and new genetic tests of function in different organisms Genes Dev 8, 133–46

Kinoshita N, Minshull J and Kirschner MW (1995) The identification of two novel ligands of the FGF receptor by a yeast screening method and their activity in Xenopus development Cell 83, 621–30

Kishigami S, Yoshikawa S, Castranio T, Okazaki K, Furuta Y and Mishina Y (2004) Bmp signaling through ACVRI is required for left–right patterning in the early mouse embryo Dev Biol 276, 185–93

Kitaguchi T, Nagai T, Nakata K, Aruga J and Mikoshiba K (2000) Zic3 is involved in the left–right specification of the Xenopus embryo Development 127, 4787–95

Kitamura K, Miura H, Miyagawa-Tomita S, Yanazawa M, Katoh-Fukui Y, Suzuki R, Ohuchi H, Suehiro A, Motegi Y, Nakahara Y et al. (1999) Mouse Pitx2 deficiency leads to anomalies of the ventral body wall, heart, extra- and periocular mesoderm and right pulmonary isomerism Development 126, 5749–58

Kosaki K, Bassi MT, Kosaki R, Lewin M, Belmont J, Schauer G and Casey B (1999a) Characterization and mutation analysis of human LEFTY A and LEFTY B, homologues of murine genes implicated in left–right axis development Am J Hum Genet 64, 712–21

Kosaki R, Gebbia M, Kosaki K, Lewin M, Bowers P, Towbin JA and Casey B (1999b) Left–right axis malformations associated with mutations in ACVR2B, the gene for human activin receptor type IIB Am J Med Genet 82, 70–6

Kothapalli R, Buyuksal I, Wu SQ, Chegini N and Tabibzadeh S (1997) Detection of ebaf, a novel human gene of the transforming growth factor beta superfamily association of gene expression with endometrial bleeding J Clin Invest 99, 2342–50

Kozminski KG, Johnson KA, Forscher P and Rosenbaum JL (1993) A motility in the eukaryotic flagellum unrelated to flagellar beating Proc Natl Acad Sci U S A 90, 5519–23

Krebs LT, Iwai N, Nonaka S, Welsh IC, Lan Y, Jiang R, Saijoh Y, O'Brien TP, Hamada H and Gridley T (2003) Notch signaling regulates left–right asymmetry determination by inducing Nodal expression Genes Dev 17, 1207–12

Kuehl KS and Loffredo C (2002) Risk factors for heart disease associated with abnormal sidedness Teratology 66, 242–8

Lai E, Prezioso VR, Smith E, Litvin O, Costa RH and Darnell JE, Jr. (1990) HNF-3A, a hepatocyte-enriched transcription factor of novel structure is regulated transcriptionally Genes Dev 4, 1427–36

Layton WM, Jr. (1976) Random determination of a developmental process: reversal of normal visceral asymmetry in the mouse J Hered 67, 336–8

Lee J, Platt KA, Censullo P and Ruiz i Altaba A (1997) Gli1 is a target of Sonic hedgehog that induces ventral neural tube development Development 124, 2537–52

Levin M (1997) Left–right asymmetry in vertebrate embryogenesis Bioessays 19, 287–96

Levin M (1998a) Left–right asymmetry and the chick embryo Semin Cell Dev Biol 9, 67–76

Levin M (1998b) The roles of activin and follistatin signaling in chick gastrulation Int J Dev Biol 42, 553–9

Levin M (2005) Left–right asymmetry in embryonic development: a comprehensive review Mech Dev 122, 3–25

Levin M, Johnson RL, Stern CD, Kuehn M and Tabin C (1995) A molecular pathway determining left–right asymmetry in chick embryogenesis Cell 82, 803–14

Levin M and Mercola M (1998a) The compulsion of chirality: toward an understanding of left–right asymmetry Genes Dev 12, 763–9

Levin M and Mercola M (1998b) Gap junctions are involved in the early generation of left–right asymmetry Dev Biol 203, 90–105

Levin M and Mercola M (1999) Gap junction-mediated transfer of left–right patterning signals in the early chick blastoderm is upstream of Shh asymmetry in the node Development 126, 4703–14

Levin M, Pagan S, Roberts DJ, Cooke J, Kuehn MR and Tabin CJ (1997) Left/right patterning signals and the independent regulation of different aspects of situs in the chick embryo Dev Biol 189, 57-67

Levin M, Roberts DJ, Holmes LB and Tabin C (1996) Laterality defects in conjoined twins Nature 384, 321

Levin M, Thorlin T, Robinson KR, Nogi T and Mercola M (2002) Asymmetries in H+/K+-ATPase and cell membrane potentials comprise a very early step in left–right patterning Cell 111, 77–89

Levy V and Khaner O (1998) Limited left–right cell migration across the midline of the gastrulating avian embryo Dev Genet 23, 175–84

Lin AE, Ticho BS, Houde K, Westgate MN and Holmes LB (2000) Heterotaxy: associated conditions and hospital-based prevalence in newborns Genet Med 2, 157–72

Litingtung Y, Lei L, Westphal H and Chiang C (1998) Sonic hedgehog is essential to foregut development Nat Genet 20, 58–61

Liu A, Wang B and Niswander LA (2005) Mouse intraflagellar transport proteins regulate both the activator and repressor functions of Gli transcription factors Development 132, 3103–11

Liu C, Liu W, Lu MF, Brown NA and Martin JF (2001) Regulation of left–right asymmetry by thresholds of Pitx2c activity Development 128, 2039–48

Liu C, Liu W, Palie J, Lu MF, Brown NA and Martin JF (2002) Pitx2c patterns anterior myocardium and aortic arch vessels and is required for local cell movement into atrioventricular cushions Development 129, 5081–91

Logan M, Pagan-Westphal SM, Smith DM, Paganessi L and Tabin CJ (1998) The transcription factor Pitx2 mediates situs-specific morphogenesis in response to left–right asymmetric signals Cell 94, 307–17

Lohr JL, Danos MC and Yost HJ (1997) Left–right asymmetry of a nodal-related gene is regulated by dorsoanterior midline structures during Xenopus development Development 124, 1465–72

Lowe LA, Supp DM, Sampath K, Yokoyama T, Wright CV, Potter SS, Overbeek P and Kuehn MR (1996) Conserved left–right asymmetry of nodal expression and alterations in murine situs inversus Nature 381, 158–61

Lu MF, Pressman C, Dyer R, Johnson RL and Martin JF (1999) Function of Rieger syndrome gene in left–right asymmetry and craniofacial development Nature 401, 276–8

Lyons I, Parsons LM, Hartley L, Li R, Andrews JE, Robb L and Harvey RP (1995) Myogenic and morphogenetic defects in the heart tubes of murine embryos lacking the homeo box gene Nkx2–5 Genes Dev 9, 1654–66

Ma Y, Erkner A, Gong R, Yao S, Taipale J, Basler K and Beachy PA (2002) Hedgehog-mediated patterning of the mammalian embryo requires transporter-like function of dispatched Cell 111, 63–75

Maclean K and Dunwoodie SL (2004) Breaking symmetry: a clinical overview of left–right patterning Clin Genet 65, 441–57

Maiti AK, Bartoloni L, Mitchison HM, Meeks M, Chung E, Spiden S, Gehrig C, Rossier C, DeLozier-Blanchet CD, Blouin J et al. (2000) No deleterious mutations in the FOXJ1 (alias HFH-4) gene in patients with primary ciliary dyskinesia (PCD) Cytogenet Cell Genet 90, 119–22

Manner J (2000) Cardiac looping in the chick embryo: a morphological review with special reference to terminological and biomechanical aspects of the looping process Anat Rec 259, 248–62

Manner J (2001) Does an equivalent of the “ventral node” exist in chick embryos? A scanning electron microscopic study Anat Embryol (Berl) 203, 481–90

Manner J (2004) On rotation, torsion, lateralization, and handedness of the embryonic heart loop: new insights from a simulation model for the heart loop of chick embryos Anat Rec A Discov Mol Cell Evol Biol 278, 481–92

Marques S, Borges AC, Silva AC, Freitas S, Cordenonsi M and Belo JA (2004) The activity of the Nodal antagonist Cerl-2 in the mouse node is required for correct L/R body axis Genes Dev 18, 2342–7

Marszalek JR and Goldstein LS (2000) Understanding the functions of kinesin-II Biochim Biophys Acta 1496, 142–50

Marszalek JR, Ruiz-Lozano P, Roberts E, Chien KR and Goldstein LS (1999) Situs inversus and embryonic ciliary morphogenesis defects in mouse mutants lacking the KIF3A subunit of kinesin-II Proc Natl Acad Sci U S A 96, 5043–8

Martinez-Frias ML, Urioste M, Bermejo E, Rodriguez-Pinilla E, Felix V, Paisan L, Martinez S, Egues J, Gomez F, Aparicio P et al. (1995) Primary midline developmental field. II. Clinical/epidemiological analysis of alteration of laterality (normal body symmetry and asymmetry) Am J Med Genet 56, 382–8

Massague J (1996) TGFbeta signaling: receptors, transducers, and Mad proteins Cell 85, 947–50

Massague J and Chen YG (2000) Controlling TGF-beta signaling Genes Dev 14, 627–44

May SR, Ashique AM, Karlen M, Wang B, Shen Y, Zarbalis K, Reiter J, Ericson J and Peterson AS (2005) Loss of the retrograde motor for IFT disrupts localization of Smo to cilia and prevents the expression of both activator and repressor functions of Gli Dev Biol 287, 378–89

McGrath J and Brueckner M (2003) Cilia are at the heart of vertebrate left–right asymmetry Curr Opin Genet Dev 13, 385–92

McGrath J, Horwich AL and Brueckner M (1992) Duplication/deficiency mapping of situs inversus viscerum (iv), a gene that determines left–right asymmetry in the mouse Genomics 14, 643–8

McGrath J, Somlo S, Makova S, Tian X and Brueckner M (2003) Two populations of node monocilia initiate left–right asymmetry in the mouse Cell 114, 61–73

McMahon JA, Takada S, Zimmerman LB, Fan CM, Harland RM and McMahon AP (1998) Noggin-mediated antagonism of BMP signaling is required for growth and patterning of the neural tube and somite Genes Dev 12, 1438–52

McMannus C (2002) Right hand, left hand. London: Weidenfeld and Nicolson

McQuinn TC, Miga DE, Mjaatvedt CH, Phelps AL and Wessels A (2001) Cardiopulmonary malformations in the inv/inv mouse Anat Rec 263, 62–71

Melloy PG, Ewart JL, Cohen MF, Desmond ME, Kuehn MR and Lo CW (1998) No turning, a mouse mutation causing left–right and axial patterning defects Dev Biol 193, 77–89

Meno C, Gritsman K, Ohishi S, Ohfuji Y, Heckscher E, Mochida K, Shimono A, Kondoh H, Talbot WS, Robertson EJ et al. (1999) Mouse Lefty2 and zebrafish antivin are feedback inhibitors of nodal signaling during vertebrate gastrulation Mol Cell 4, 287–98

Meno C, Ito Y, Saijoh Y, Matsuda Y, Tashiro K, Kuhara S and Hamada H (1997) Two closely-related left–right asymmetrically expressed genes, lefty-1 and lefty-2: their distinct expression domains, chromosomal linkage and direct neuralizing activity in Xenopus embryos Genes Cells 2, 513–24

Meno C, Saijoh Y, Fujii H, Ikeda M, Yokoyama T, Yokoyama M, Toyoda Y and Hamada H (1996) Left–right asymmetric expression of the TGF beta-family member lefty in mouse embryos Nature 381, 151–5

Meno C, Shimono A, Saijoh Y, Yashiro K, Mochida K, Ohishi S, Noji S, Kondoh H and Hamada H (1998) lefty-1 is required for left–right determination as a regulator of lefty-2 and nodal Cell 94, 287–97

Meno C, Takeuchi J, Sakuma R, Koshiba-Takeuchi K, Ohishi S, Saijoh Y, Miyazaki J, ten Dijke P, Ogura T and Hamada H (2001) Diffusion of nodal signaling activity in the absence of the feedback inhibitor Lefty2 Dev Cell 1, 127–38

Metzler M, Gertz A, Sarkar M, Schachter H, Schrader JW and Marth JD (1994) Complex asparagine-linked oligosaccharides are required for morphogenic events during post-implantation development Embo J 13, 2056–65

Meyers EN and Martin GR (1999) Differences in left–right axis pathways in mouse and chick: functions of FGF8 and SHH Science 285, 403–6

Mikami A, Tynan SH, Hama T, Luby-Phelps K, Saito T, Crandall JE, Besharse JC and Vallee RB (2002) Molecular structure of cytoplasmic dynein 2 and its distribution in neuronal and ciliated cells J Cell Sci 115, 4801–8

Mishina Y (2003) Function of bone morphogenetic protein signaling during mouse development Front Biosci 8, d855–69

Mochizuki T, Saijoh Y, Tsuchiya K, Shirayoshi Y, Takai S, Taya C, Yonekawa H, Yamada K, Nihei H, Nakatsuji N et al. (1998) Cloning of inv, a gene that controls left/right asymmetry and kidney development Nature 395, 177–81

Monsoro-Burq A and Le Douarin N (2000) Left–right asymmetry in BMP4 signalling pathway during chick gastrulation Mech Dev 97, 105–8

Monsoro-Burq A and Le Douarin NM (2001) BMP4 plays a key role in left–right patterning in chick embryos by maintaining Sonic Hedgehog asymmetry Mol Cell 7, 789–99

Moore KL and Persaud TVN (1998) The Developing Human: Clinically Oriented Embryology. Philadelphia: W.B. Saunders Company

Morgan D, Goodship J, Essner JJ, Vogan KJ, Turnpenny L, Yost HJ, Tabin CJ and Strachan T (2002) The left–right determinant inversin has highly conserved ankyrin repeat and IQ domains and interacts with calmodulin Hum Genet 110, 377–84

Morgan D, Turnpenny L, Goodship J, Dai W, Majumder K, Matthews L, Gardner A, Schuster G, Vien L, Harrison W et al. (1998) Inversin, a novel gene in the vertebrate left–right axis pathway, is partially deleted in the inv mouse Nat Genet 20, 149–56

Motoyama J, Liu J, Mo R, Ding Q, Post M and Hui CC (1998) Essential function of Gli2 and Gli3 in the formation of lung, trachea and oesophagus Nat Genet 20, 54–7

Moyer JH, Lee-Tischler MJ, Kwon HY, Schrick JJ, Avner ED, Sweeney WE, Godfrey VL, Cacheiro NL, Wilkinson JE and Woychik RP (1994) Candidate gene associated with a mutation causing recessive polycystic kidney disease in mice Science 264, 1329–33

Muller F and O'Rahilly R (1987) The development of the human brain, the closure of the caudal neuropore, and the beginning of secondary neurulation at stage 12 Anat Embryol (Berl) 176, 413–30

Murcia NS, Richards WG, Yoder BK, Mucenski ML, Dunlap JR and Woychik RP (2000) The Oak Ridge Polycystic Kidney (orpk) disease gene is required for left–right axis determination Development 127, 2347–55

Nakaya MA, Biris K, Tsukiyama T, Jaime S, Rawls JA and Yamaguchi TP (2005) Wnt3alinks left–right determination with segmentation and anteroposterior axis elongation Development 132, 5425–36

Nanni L, Ming JE, Bocian M, Steinhaus K, Bianchi DW, Die-Smulders C, Giannotti A, Imaizumi K, Jones KL, Campo MD et al. (1999) The mutational spectrum of the sonic hedgehog gene in holoprosencephaly: SHH mutations cause a significant proportion of autosomal dominant holoprosencephaly Hum Mol Genet 8, 2479–88

Nascone N and Mercola M (1997) Organizer induction determines left–right asymmetry in Xenopus Dev Biol 189, 68–78

Nauli SM, Alenghat FJ, Luo Y, Williams E, Vassilev P, Li X, Elia AE, Lu W, Brown EM, Quinn SJ et al. (2003) Polycystins 1 and 2 mediate mechanosensation in the primary cilium of kidney cells Nat Genet 33, 129–37

Neesen J, Drenckhahn JD, Tiede S, Burfeind P, Grzmil M, Konietzko J, Dixkens C, Kreutzberger J, Laccone F and Omran H (2002) Identification of the human ortholog of the t-complex-encoded protein TCTE3 and evaluation as a candidate gene for primary ciliary dyskinesia Cytogenet Genome Res 98, 38–44

Niederreither K, Vermot J, Messaddeq N, Schuhbaur B, Chambon P and Dolle P (2001) Embryonic retinoic acid synthesis is essential for heart morphogenesis in the mouse Development 128, 1019–31

Nielsen C, Murtaugh LC, Chyung JC, Lassar A and Roberts DJ (2001) Gizzard formation and the role of Bapx1 Dev Biol 231, 164–74

Nomura M and Li E (1998) Smad2 role in mesoderm formation, left–right patterning and craniofacial development Nature 393, 786–90

Nonaka S, Shiratori H, Saijoh Y and Hamada H (2002) Determination of left–right patterning of the mouse embryo by artificial nodal flow Nature 418, 96–9

Nonaka S, Tanaka Y, Okada Y, Takeda S, Harada A, Kanai Y, Kido M and Hirokawa N (1998) Randomization of left–right asymmetry due to loss of nodal cilia generating leftward flow of extraembryonic fluid in mice lacking KIF3B motor protein Cell 95, 829–37

Norris DP, Brennan J, Bikoff EK and Robertson EJ (2002) The Foxh1-dependent autoregulatory enhancer controls the level of Nodal signals in the mouse embryo Development 129, 3455–68

Norris DP and Robertson EJ (1999) Asymmetric and node-specific nodal expression patterns are controlled by two distinct cis-acting regulatory elements Genes Dev 13, 1575–88

Nugent WC, Niles NW, 2nd, Schults W, Plume SK, Wolf BH, Tarbox GH, Robb JF and Nelson EC (1994) Increasing the value of cardiac care: the Dartmouth approach Qual Lett Healthc Lead 6, 53–7

Nurnberger J, Kavapurackal R, Zhang SJ, Saez AO, Heusch G, Philipp T, Pietruck F and Kribben A (2006) Differential tissue distribution of the Invs gene product inversin Cell Tissue Res 323, 147–55

Oh SP and Li E (1997) The signaling pathway mediated by the type IIB activin receptor controls axial patterning and lateral asymmetry in the mouse Genes Dev 11, 1812–26

Oh SP, Yeo CY, Lee Y, Schrewe H, Whitman M and Li E (2002) Activin type IIA and IIB receptors mediate Gdf11 signaling in axial vertebral patterning Genes Dev 16, 2749–54

Okada Y, Nonaka S, Tanaka Y, Saijoh Y, Hamada H and Hirokawa N (1999) Abnormal nodal flow precedes situs inversus in iv and inv mice Mol Cell 4, 459–68

Okada Y, Takeda S, Tanaka Y, Belmonte JC and Hirokawa N (2005) Mechanism of nodal flow: a conserved symmetry breaking event in left–right axis determination Cell 121, 633–44

Olbrich H, Haffner K, Kispert A, Volkel A, Volz A, Sasmaz G, Reinhardt R, Hennig S, Lehrach H, Konietzko N et al. (2002) Mutations in DNAH5 cause primary ciliary dyskinesia and randomization of left–right asymmetry Nat Genet 30, 143–4

Otto EA, Schermer B, Obara T, O'Toole JF, Hiller KS, Mueller AM, Ruf RG, Hoefele J, Beekmann F, Landau D et al. (2003) Mutations in INVS encoding inversin cause nephronophthisis type 2, linking renal cystic disease to the function of primary cilia and left–right axis determination Nat Genet 34, 413–20

Pagan-Westphal SM and Tabin CJ (1998) The transfer of left–right positional information during chick embryogenesis Cell 93, 25–35

Palmer AR (2004) Symmetry breaking and the evolution of development Science 306, 828–33

Patel K, Isaac A and Cooke J (1999) Nodal signalling and the roles of the transcription factors SnR and Pitx2 in vertebrate left–right asymmetry Curr Biol 9, 609–12

Patterson KD, Drysdale TA and Krieg PA (2000) Embryonic origins of spleen asymmetry Development 127, 167–75

Pazour GJ, Agrin N, Walker BL and Witman GB (2006) Identification of predicted human outer dynein arm genes: candidates for primary ciliary dyskinesia genes J Med Genet 43, 62–73

Pearce JJ, Penny G and Rossant J (1999) A mouse cerberus/Dan-related gene family Dev Biol 209, 98–110

Pennarun G, Chapelin C, Escudier E, Bridoux AM, Dastot F, Cacheux V, Goossens M, Amselem S and Duriez B (2000) The human dynein intermediate chain 2 gene (DNAI2): cloning, mapping, expression pattern, and evaluation as a candidate for primary ciliary dyskinesia Hum Genet 107, 642–9

Pennekamp P, Karcher C, Fischer A, Schweickert A, Skryabin B, Horst J, Blum M and Dworniczak B (2002) The ion channel polycystin-2 is required for left–right axis determination in mice Curr Biol 12, 938–43

Perrone CA, Yang P, O'Toole E, Sale WS and Porter ME (1998) The Chlamydomonas IDA7 locus encodes a 140-kDa dynein intermediate chain required to assemble the I1 inner arm complex Mol Biol Cell 9, 3351–65

Piedra ME, Icardo JM, Albajar M, Rodriguez-Rey JC and Ros MA (1998) Pitx2 participates in the late phase of the pathway controlling left–right asymmetry Cell 94, 319–24

Piedra ME and Ros MA (2002) BMP signaling positively regulates Nodal expression during left right specification in the chick embryo Development 129, 3431–40

Praetorius HA and Spring KR (2001) Bending the MDCK cell primary cilium increases intracellular calcium J Membr Biol 184, 71–9

Przemeck GK, Heinzmann U, Beckers J and Hrabe de Angelis M (2003) Node and midline defects are associated with left–right development in Delta1 mutant embryos Development 130, 3–13

Purandare SM, Ware SM, Kwan KM, Gebbia M, Bassi MT, Deng JM, Vogel H, Behringer RR, Belmont JW and Casey B (2002) A complex syndrome of left–right axis, central nervous system and axial skeleton defects in Zic3 mutant mice Development 129, 2293–302

Qiu D, Cheng SM, Wozniak L, McSweeney M, Perrone E and Levin M (2005) Localization and loss-of-function implicates ciliary proteins in early, cytoplasmic roles in left–right asymmetry Dev Dyn 234, 176–89

Quaderi NA, Schweiger S, Gaudenz K, Franco B, Rugarli EI, Berger W, Feldman GJ, Volta M, Andolfi G, Gilgenkrantz S et al. (1997) Opitz G/BBB syndrome, a defect of midline development, is due to mutations in a new RING finger gene on Xp22 Nat Genet 17, 285–91

Ramsdell AF (2005) Left–right asymmetry and congenital cardiac defects: getting to the heart of the matter in vertebrate left–right axis determination Dev Biol 288, 1–20

Ramsdell AF and Yost HJ (1999) Cardiac looping and the vertebrate left–right axis: antagonism of left-sided Vg1 activity by a right-sided ALK2-dependent BMP pathway Development 126, 5195–205

Rankin CT, Bunton T, Lawler AM and Lee SJ (2000) Regulation of left–right patterning in mice by growth/differentiation factor-1 Nat Genet 24, 262–5

Raya A and Belmonte JC (2006) Left–right asymmetry in the vertebrate embryo: from early information to higher-level integration Nat Rev Genet 7, 283–93

Raya A, Kawakami Y, Rodriguez-Esteban C, Ibanes M, Rasskin-Gutman D, Rodriguez-Leon J, Buscher D, Feijo JA and Izpisua Belmonte JC (2004) Notch activity acts as a sensor for extracellular calcium during vertebrate left–right determination Nature 427, 121–8

Raya A, Koth CM, Buscher D, Kawakami Y, Itoh T, Raya RM, Sternik G, Tsai HJ, Rodriguez-Esteban C and Izpisua-Belmonte JC (2003) Activation of Notch signaling pathway precedes heart regeneration in zebrafish Proc Natl Acad Sci U S A 100 Suppl 1, 11889–95

Reaume AG, de Sousa PA, Kulkarni S, Langille BL, Zhu D, Davies TC, Juneja SC, Kidder GM and Rossant J (1995) Cardiac malformation in neonatal mice lacking connexin43 Science 267, 1831–4

Rebagliati MR, Toyama R, Fricke C, Haffter P and Dawid IB (1998) Zebrafish nodal-related genes are implicated in axial patterning and establishing left–right asymmetry Dev Biol 199, 261–72

Rieger H (1935) Dysgenesis mesodermaiis coreneal et iridis. Z. Augenheik 86, 333

Rodriguez-Esteban C, Capdevila J, Economides AN, Pascual J, Ortiz A and Izpisua Belmonte JC (1999) The novel Cer-like protein Caronte mediates the establishment of embryonic left–right asymmetry Nature 401, 243–51

Rodriguez-Esteban C, Capdevila J, Kawakami Y and Izpisua Belmonte JC (2001) Wnt signaling and PKA control Nodal expression and left–right determination in the chick embryo Development 128, 3189–95

Roebroek AJ, Umans L, Pauli IG, Robertson EJ, van Leuven F, Van de Ven WJ and Constam DB (1998) Failure of ventral closure and axial rotation in embryos lacking the proprotein convertase Furin Development 125, 4863–76

Roessler E, Belloni E, Gaudenz K, Vargas F, Scherer SW, Tsui LC and Muenke M (1997) Mutations in the C-terminal domain of Sonic Hedgehog cause holoprosencephaly Hum Mol Genet 6, 1847–53

Romanoff AL. (1960). The Avian Embryo. New York: Macmillan

Rosenbaum JL and Witman GB (2002) Intraflagellar transport Nat Rev Mol Cell Biol 3, 813–25

Ruiz i Altaba A (1998) Combinatorial Gli gene function in floor plate and neuronal inductions by Sonic hedgehog Development 125, 2203–12

Ruscazio M, Van Praagh S, Marrass AR, Catani G, Iliceto S and Van Praagh R (1998) Interrupted inferior vena cava in asplenia syndrome and a review of the hereditary patterns of visceral situs abnormalities Am J Cardiol 81, 111–6

Ryan AK, Blumberg B, Rodriguez-Esteban C, Yonei-Tamura S, Tamura K, Tsukui T, de la Pena J, Sabbagh W, Greenwald J, Choe S et al. (1998) Pitx2 determines left–right asymmetry of internal organs in vertebrates Nature 394, 545–51

Sadler W (2004) Langman's Medical Embryology. Baltimore: Williams and Wilkins

Saijoh Y, Adachi H, Mochida K, Ohishi S, Hirao A and Hamada H (1999) Distinct transcriptional regulatory mechanisms underlie left–right asymmetric expression of lefty-1 and lefty-2 Genes Dev 13, 259–69

Saijoh Y, Adachi H, Sakuma R, Yeo CY, Yashiro K, Watanabe M, Hashiguchi H, Mochida K, Ohishi S, Kawabata M et al. (2000) Left–right asymmetric expression of lefty2 and nodal is induced by a signaling pathway that includes the transcription factor FAST2 Mol Cell 5, 35–47

Saijoh Y, Oki S, Ohishi S and Hamada H (2003) Left–right patterning of the mouse lateral plate requires nodal produced in the node Dev Biol 256, 160–72

Saijoh Y, Oki S, Tanaka C, Nakamura T, Adachi H, Yan YT, Shen MM and Hamada H (2005) Two nodal-responsive enhancers control left–right asymmetric expression of Nodal Dev Dyn 232, 1031–6

Sakuma R, Ohnishi Yi Y, Meno C, Fujii H, Juan H, Takeuchi J, Ogura T, Li E, Miyazono K and Hamada H (2002) Inhibition of Nodal signalling by Lefty mediated through interaction with common receptors and efficient diffusion Genes Cells 7, 401–12

Saloman DS, Bianco C, Ebert AD, Khan NI, De Santis M, Normanno N, Wechselberger C, Seno M, Williams K, Sanicola M et al. (2000) The EGF-CFC family: novel epidermal growth factor-related proteins in development and cancer Endocr Relat Cancer 7, 199–226

Sampath K, Rubinstein AL, Cheng AM, Liang JO, Fekany K, Solnica-Krezel L, Korzh V, Halpern ME and Wright CV (1998) Induction of the zebrafish ventral brain and floorplate requires cyclops/nodal signalling Nature 395, 185–9

Sarmah B, Latimer AJ, Appel B and Wente SR (2005) Inositol polyphosphates regulate zebrafish left–right asymmetry Dev Cell 9, 133–45

Sasaki H and Hogan BL (1994) HNF-3 beta as a regulator of floor plate development Cell 76, 103–15

Sasaki H, Hui C, Nakafuku M and Kondoh H (1997) A binding site for Gli proteins is essential for HNF-3beta floor plate enhancer activity in transgenics and can respond to Shh in vitro Development 124, 1313–22

Sasaki R, Minowada J, Bollum FJ and Miura Y (1990) Polyadenylic acid polymerase activity in chronic myelogenous leukemia Leuk Res 14, 273–8

Schier AF (2003) Nodal signaling in vertebrate development Annu Rev Cell Dev Biol 19, 589–621

Schier AF and Shen MM (2000) Nodal signalling in vertebrate development Nature 403, 385–9

Schlange T, Arnold HH and Brand T (2002) BMP2 is a positive regulator of Nodal signaling during left–right axis formation in the chicken embryo Development 129, 3421–9

Schlange T, Schnipkoweit I, Andree B, Ebert A, Zile MH, Arnold HH and Brand T (2001) Chick CFC controls Lefty1 expression in the embryonic midline and nodal expression in the lateral plate Dev Biol 234, 376–89

Schneider A, Mijalski T, Schlange T, Dai W, Overbeek P, Arnold HH and Brand T (1999) The homeobox gene NKX3.2 is a target of left–right signalling and is expressed on opposite sides in chick and mouse embryos Curr Biol 9, 911–4

Schweickert A, Campione M, Steinbeisser H and Blum M (2000) Pitx2 isoforms: involvement of Pitx2c but not Pitx2a or Pitx2b in vertebrate left–right asymmetry Mech Dev 90, 41–51

Sefton M, Sanchez S and Nieto MA (1998) Conserved and divergent roles for members of the Snail family of transcription factors in the chick and mouse embryo Development 125, 3111–21

Semina EV, Reiter R, Leysens NJ, Alward WL, Small KW, Datson NA, Siegel-Bartelt J, Bierke-Nelson D, Bitoun P, Zabel BU et al. (1996) Cloning and characterization of a novel bicoid-related homeobox transcription factor gene RIEG, involved in Rieger syndrome Nat Genet 14, 392–9

Shamim H and Mason I (1999) Expression of Fgf4 during early development of the chick embryo Mech Dev 85, 189–92

Shen MM, Wang H and Leder P (1997) A differential display strategy identifies Cryptic, a novel EGF-related gene expressed in the axial and lateral mesoderm during mouse gastrulation Development 124, 429–42

Shenefelt RE (1972) Morphogenesis of malformations in hamsters caused by retinoic acid: relation to dose and stage at treatment Teratology 5, 103–18

Shi Y and Massague J (2003) Mechanisms of TGF-beta signaling from cell membrane to the nucleus Cell 113, 685–700

Shiratori H, Sakuma R, Watanabe M, Hashiguchi H, Mochida K, Sakai Y, Nishino J, Saijoh Y, Whitman M and Hamada H (2001) Two-step regulation of left–right asymmetric expression of Pitx2: initiation by nodal signaling and maintenance by Nkx2 Mol Cell 7, 137–49

Simons M, Gloy J, Ganner A, Bullerkotte A, Bashkurov M, Kronig C, Schermer B, Benzing T, Cabello OA, Jenny A et al. (2005) Inversin, the gene product mutated in nephronophthisis type II, functions as a molecular switch between Wnt signaling pathways Nat Genet 37, 537–43

Skandalakis and Gray (1994) Embryology for Surgeons: Williams and Wilkins

Smith SM, Dickman ED, Thompson RP, Sinning AR, Wunsch AM and Markwald RR (1997) Retinoic acid directs cardiac laterality and the expression of early markers of precardiac asymmetry Dev Biol 182, 162–71

Snell GD and Stevens LC. (1966). Early Embryology. In biology of the laboratory mouse. New York: McGraw-Hill

Solloway MJ and Robertson EJ (1999) Early embryonic lethality in Bmp5;Bmp7 double mutant mice suggests functional redundancy within the 60A subgroup Development 126, 1753–68

Song J, Oh SP, Schrewe H, Nomura M, Lei H, Okano M, Gridley T and Li E (1999) The type II activin receptors are essential for egg cylinder growth, gastrulation, and rostral head development in mice Dev Biol 213, 157–69

Spemann H and Falkenberg H (1919) Über Asymmetrische Entwicklung und Situs inversus viscerum bei Zwillingen und Doppelbildungen Wilhelm Roux' Archives 45, 371–422

Splitt MP and Goodship JA (1997) Another case of maternal uniparental disomy chromosome 14 syndrome Am J Med Genet 72, 239–40

Sternick EB, Marcio Gerken L, Max R and Osvaldo Vrandecic M (2004) Radiofrequency catheter ablation of an accessory pathway in a patient with Wolff-Parkinson-White and Kartagener's syndrome Pacing Clin Electrophysiol 27, 401–4

Streit A, Lee KJ, Woo I, Roberts C, Jessell TM and Stern CD (1998) Chordin regulates primitive streak development and the stability of induced neural cells, but is not sufficient for neural induction in the chick embryo Development 125, 507–19

Sulik K, Dehart DB, Iangaki T, Carson JL, Vrablic T, Gesteland K and Schoenwolf GC (1994) Morphogenesis of the murine node and notochordal plate Dev Dyn 201, 260–78

Supp DM, Brueckner M, Kuehn MR, Witte DP, Lowe LA, McGrath J, Corrales J and Potter SS (1999) Targeted deletion of the ATP binding domain of left–right dynein confirms its role in specifying development of left–right asymmetries Development 126, 5495–504

Supp DM, Witte DP, Potter SS and Brueckner M (1997) Mutation of an axonemal dynein affects left–right asymmetry in inversus viscerum mice Nature 389, 963–6

Tabin C (2005) Do we know anything about how left–right asymmetry is first established in the vertebrate embryo? J Mol Histol 36, 317–23

Tabin CJ and Vogan KJ (2003) A two-cilia model for vertebrate left–right axis specification Genes Dev 17, 1–6

Takeda S, Yonekawa Y, Tanaka Y, Okada Y, Nonaka S and Hirokawa N (1999) Left–right asymmetry and kinesin superfamily protein KIF3A: new insights in determination of laterality and mesoderm induction by kif3A-/- mice analysis J Cell Biol 145, 825–36

Tanaka Y, Okada Y and Hirokawa N (2005) FGF-induced vesicular release of Sonic hedgehog and retinoic acid in leftward nodal flow is critical for left–right determination Nature 435, 172–7

Theiler K. (1989). The Mouse House: Atlas of Mouse Development. New York: Springer-Verlag

Thisse C and Thisse B (1999) Antivin, a novel and divergent member of the TGFbeta superfamily, negatively regulates mesoderm induction Development 126, 229–40

Ticho BS, Goldstein AM and Van Praagh R (2000) Extracardiac anomalies in the heterotaxy syndromes with focus on anomalies of midline-associated structures Am J Cardiol 85, 729–34

Torgersen J (1950) Situs inversus, asymmetry, and twinning Am. J. Hum. Gent. 2, 361–370

Tribioli C and Lufkin T (1997) Molecular cloning, chromosomal mapping and developmental expression of BAPX1, a novel human homeobox-containing gene homologous to Drosophila bagpipe Gene 203, 225–33

Tsuda T, Majumder K and Linask KK (1998) Differential expression of flectin in the extracellular matrix and left–right asymmetry in mouse embryonic heart during looping stages Dev Genet 23, 203–14

Tsuda T, Philp N, Zile MH and Linask KK (1996) Left–right asymmetric localization of flectin in the extracellular matrix during heart looping Dev Biol 173, 39–50

Tsukui T, Capdevila J, Tamura K, Ruiz-Lozano P, Rodriguez-Esteban C, Yonei-Tamura S, Magallon J, Chandraratna RA, Chien K, Blumberg B et al. (1999) Multiple left–right asymmetry defects in Shh(-/-) mutant mice unveil a convergence of the shh and retinoic acid pathways in the control of Lefty-1 Proc Natl Acad Sci U S A 96, 11376–81

Twal W, Roze L and Zile MH (1995) Anti-retinoic acid monoclonal antibody localizes all-trans-retinoic acid in target cells and blocks normal development in early quail embryo Dev Biol 168, 225–34

Vallee RB and Hook P (2003) Molecular motors: A magnificent machine Nature 421, 701–2

van der Hoeven F, Schimmang T, Volkmann A, Mattei MG, Kyewski B and Ruther U (1994) Programmed cell death is affected in the novel mouse mutant Fused toes (Ft) Development 120, 2601–7

Van's Gravesande KS and Omran H (2005) Primary ciliary dyskinesia: clinical presentation, diagnosis and genetics Ann Med 37, 439–49

Vermot J, Gallego Llamas J, Fraulob V, Niederreither K, Chambon P and Dolle P (2005) Retinoic acid controls the bilateral symmetry of somite formation in the mouse embryo Science 308, 563–6

Vermot J and Pourquie O (2005) Retinoic acid coordinates somitogenesis and left-right patterning in vertebrate embryos Nature 435, 215–20

Vogan KJ and Tabin CJ (1999) A new spin on handed asymmetry Nature 397, 295, 297–8

Volkmann A, Doffinger R, Ruther U and Kyewski BA (1996) Insertional mutagenesis affecting programmed cell death leads to thymic hyperplasia and altered thymopoiesis J Immunol 156, 136–45

Wagner MK and Yost HJ (2000) Left-right development: the roles of nodal cilia Curr Biol 10 R149–51

Waldo KL, Hutson MR, Stadt HA, Zdanowicz M, Zdanowicz J and Kirby ML (2005a) Cardiac neural crest is necessary for normal addition of the myocardium to the arterial pole from the secondary heart field Dev Biol 281, 66–77

Waldo KL, Hutson MR, Ward CC, Zdanowicz M, Stadt HA, Kumiski D, Abu-Issa R and Kirby ML (2005b) Secondary heart field contributes myocardium and smooth muscle to the arterial pole of the developing heart Dev Biol 281, 78–90

Waldo KL, Kumiski DH, Wallis KT, Stadt HA, Hutson MR, Platt DH and Kirby ML (2001) Conotruncal myocardium arises from a secondary heart field Development 128, 3179–88

Waldrip WR, Bikoff EK, Hoodless PA, Wrana JL and Robertson EJ (1998) Smad2 signaling in extraembryonic tissues determines anterior-posterior polarity of the early mouse embryo Cell 92, 797–808

Wall NA, Craig EJ, Labosky PA and Kessler DS (2000) Mesendoderm induction and reversal of left-right pattern by mouse Gdf1, a Vg1-related gene Dev Biol 227, 495–509

Wang S, Yu X, Zhang T, Zhang X, Zhang Z and Chen Y (2004) Chick Pcl2 regulates the left-right asymmetry by repressing Shh expression in Hensen's node Development 131, 4381-91

Ware SM, Peng J, Zhu L, Fernbach S, Colicos S, Casey B, Towbin J and Belmont JW (2004) Identification and functional analysis of ZIC3 mutations in heterotaxy and related congenital heart defects Am J Hum Genet 74, 93–105

Wasiak S and Lohnes D (1999) Retinoic acid affects left-right patterning Dev Biol 215, 332–42

Watanabe D, Saijoh Y, Nonaka S, Sasaki G, Ikawa Y, Yokoyama T and Hamada H (2003) The left-right determinant Inversin is a component of node monocilia and other 9+0 cilia Development 130, 1725–34

Watanabe Y, Benson DW, Yano S, Akagi T, Yoshino M and Murray JC (2002) Two novel frameshift mutations in NKX2.5 result in novel features including visceral inversus and sinus venosus type ASD J Med Genet 39, 807–11

Weinstein DC, Ruiz i Altaba A, Chen WS, Hoodless P, Prezioso VR, Jessell TM and Darnell JE, Jr. (1994) The winged-helix transcription factor HNF-3 beta is required for notochord development in the mouse embryo Cell 78, 575–88

Weninger WJ, Floro KL, Bennett MB, Withington SL, Preis JI, Barbera JP, Mohun TJ and Dunwoodie SL (2005) Cited2 is required both for heart morphogenesis and establishment of the left–right axis in mouse development Development 132, 1337–48

Whitman M and Mercola M (2001) TGF-beta superfamily signaling and left–right asymmetry Sci STKE 2001 RE1

Winnier G, Blessing M, Labosky PA and Hogan BL (1995) Bone morphogenetic protein-4 is required for mesoderm formation and patterning in the mouse Genes Dev 9, 2105–16

Xie J, Murone M, Luoh SM, Ryan A, Gu Q, Zhang C, Bonifas JM, Lam CW, Hynes M, Goddard A et al. (1998) Activating Smoothened mutations in sporadic basal-cell carcinoma Nature 391, 90–2

Ya J, Erdtsieck-Ernste EB, de Boer PA, van Kempen MJ, Jongsma H, Gros D, Moorman AF and Lamers WH (1998) Heart defects in connexin43-deficient mice Circ Res 82, 360–6

Yamamoto M, Meno C, Sakai Y, Shiratori H, Mochida K, Ikawa Y, Saijoh Y and Hamada H (2001) The transcription factor FoxH1 (FAST) mediates Nodal signaling during anterior-posterior patterning and node formation in the mouse Genes Dev 15, 1242–56

Yamamoto M, Mine N, Mochida K, Sakai Y, Saijoh Y, Meno C and Hamada H (2003) Nodal signaling induces the midline barrier by activating Nodal expression in the lateral plate Development 130, 1795–804

Yan YT, Gritsman K, Ding J, Burdine RD, Corrales JD, Price SM, Talbot WS, Schier AF and Shen MM (1999) Conserved requirement for EGF-CFC genes in vertebrate left–right axis formation Genes Dev 13, 2527–37

Yeo C and Whitman M (2001) Nodal signals to Smads through Cripto-dependent and Cripto-independent mechanisms Mol Cell 7, 949–57

Yokouchi Y, Vogan KJ, Pearse RV, 2nd and Tabin CJ (1999) Antagonistic signaling by Caronte, a novel Cerberus-related gene, establishes left–right asymmetric gene expression Cell 98, 573–83

Yokoyama T, Copeland NG, Jenkins NA, Montgomery CA, Elder FF and Overbeek PA (1993) Reversal of left–right asymmetry: a situs inversus mutation Science 260, 679–82

Yoshioka H, Meno C, Koshiba K, Sugihara M, Itoh H, Ishimaru Y, Inoue T, Ohuchi H, Semina EV, Murray JC et al. (1998) Pitx2, a bicoid-type homeobox gene, is involved in a lefty-signaling pathway in determination of left–right asymmetry Cell 94, 299–305

Yost HJ (1998) Left–right development in Xenopus and zebrafish Semin Cell Dev Biol 9, 61–6

Yost HJ (2001) Establishment of left–right asymmetry Int Rev Cytol 203, 357–81

Zhang J, Talbot WS and Schier AF (1998) Positional cloning identifies zebrafish one-eyed pinhead as a permissive EGF-related ligand required during gastrulation Cell 92, 241–51

Zhang XM, Ramalho-Santos M and McMahon AP (2001) Smoothened mutants reveal redundant roles for Shh and Ihh signaling including regulation of L/R symmetry by the mouse node Cell 106, 781–92

Zhang YJ, O'Neal WK, Randell SH, Blackburn K, Moyer MB, Boucher RC and Ostrowski LE (2002) Identification of dynein heavy chain 7 as an inner arm component of human cilia that is synthesized but not assembled in a case of primary ciliary dyskinesia J Biol Chem 277, 17906–15

Zhou X, Sasaki H, Lowe L, Hogan BL and Kuehn MR (1993) Nodal is a novel TGF-beta-like gene expressed in the mouse node during gastrulation Nature 361, 543–7

Zhu L, Belmont JW and Ware SM (2006) Genetics of human heterotaxias Eur J Hum Genet 14, 17–25

Zhu L, Marvin MJ, Gardiner A, Lassar AB, Mercola M, Stern CD and Levin M (1999) Cerberus regulates left–right asymmetry of the embryonic head and heart In Curr Biol, vol. 9 (ed.) pp. 931–8

1
Appendix

A.1
Ciliary Structure and Function

Cilia are membrane projections from eukaryotic cells that contain doublets of parallel microtubules and serve two important functions: motility and sensory reception (Figs. 16 and 17). Flagella have the same structure as cilia but they are longer and usually appear singly in each cell. Cilia are composed of an outer membrane embracing the axoneme. Most cilia have nine pairs of microtubule doublets and two microtubule singlets in the center. This organization is usually referred to as 9+2. Each doublet consists of an A and a B tubule, made up of 13 and 11 tubulin units, respectively. The tubule A of the peripheral doublets is joined to the central microtubules by radial spokes. Adjacent doublets are joined by Nexin, a highly elastic protein, and the two singlets in the center of the axoneme are joined by a connecting bridge. Finally, each tubule A bears two arms, an inner and an outer arm, made of the protein dynein, that connect it with the next tubule (Fig. 16). The dynein arms are responsible for the sliding of the tubules, therefore for the ciliary movement. In their absence cilia become rigid structures.

In certain types of cilia the axoneme lacks the central pair of tubules and consequently these cilia are referred to as 9+0. These cilia have been called "primary cilia" and are usually found in the cell as a solitary cilium, or monocilium. The axoneme also contains a number of multiprotein complexes that interconnect the different components. At the base of the cilium is the microtubule-organizing center, or basal body, consisting of the centriole. The region connecting the basal body and the axoneme is the transition zone.

Traditionally, two types of cilia have been distinguished according to their motility and structure. Motile cilia constantly beat in one direction and exhibit the 9+2 structure. Monocilia, with the 9+0 structure, were previously considered to be immotile and to function as sensors. In humans, for instance, motile cilia are found in the lining of the trachea and respiratory airways where they brush mucus and dirt out of the lungs, and in the oviducts where they move the ovum from the ovary to the uterus. Examples of monociliated cells are most embryonic cells and the epithelial cells of kidney tubules (Fig. 17). Recently, however, it have been demonstrated that the monocilia of the node, which are primary cilia with the 9+0 structure, have a rapid rotational movement. Accordingly Ibanez-Tallon et al. (2003) have proposed classifying cilia into four subtypes: (1) motile 9+2 cilia (e.g., respiratory cilia), (2) motile 9+0 cilia (e.g., nodal cilia), (3) sensory 9+2 cilia (e.g., vestibulary cilia), and (4) sensory 9+0 cilia (renal monocilia and photoreceptor connective cilia). Defects in cilia are at the root of several human diseases; the best known is primary ciliary dyskinesia (PCD). Defects in motile cilia produce respiratory diseases, sterility in males, and subfertility in females. Defects in monocilia are involved in the pathogenesis of polycystic kidney disease, retinal degeneration, and liver fibrosis.

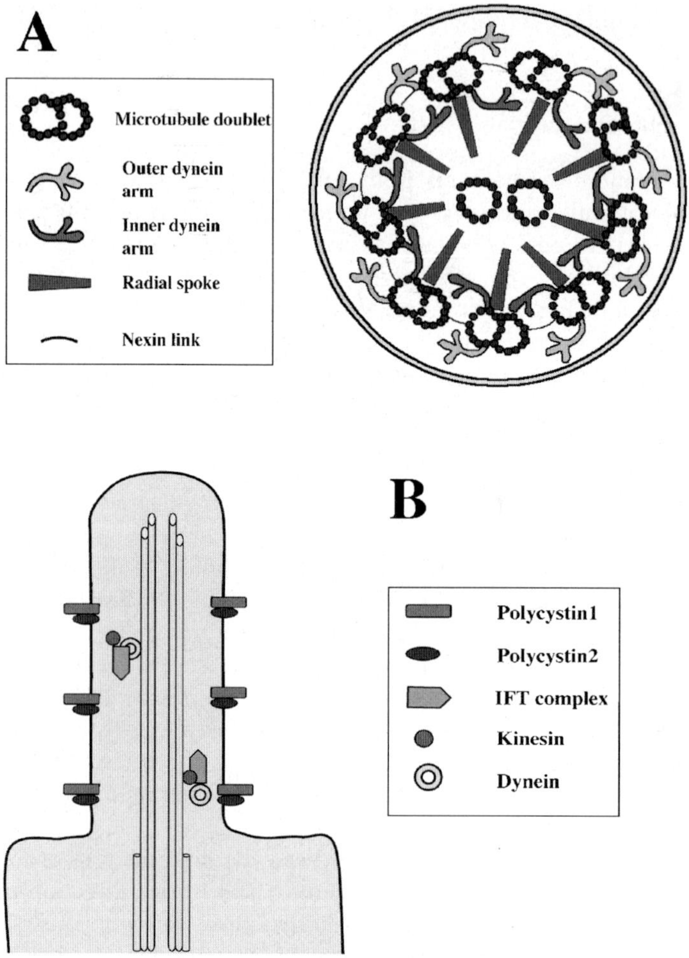

Fig. 16A, B Schematic diagram of the ciliary structure. **A** Diagrammatic representation of a transverse section through a 9+2 cilium. The different components are represented as explained in the *side box*. **B** Diagrammatic representation of a longitudinal section through a cilium to illustrate the intraflagellar transport. The different components are represented as explained in the *side box*

Dyneins are essential structural components of cilia. Dyneins are motor molecules that have been classified as cytoplasmic or axonemal. Cytoplasmic dyneins are involved in cytoplasmic trafficking of vesicles, retrograde axonemal transport, and spindle organization during mitosis, whereas axonemal dyneins move cilia and flagella by sliding adjacent microtubules. However, cytoplasmic dynein 2 is specific to ciliated cells, where it is involved in axonemal transport (Mikami et al. 2002; Vallee and Hook 2003). Dyneins form very large multisubunit complexes containing two or three heavy chains, which include the motor domain,

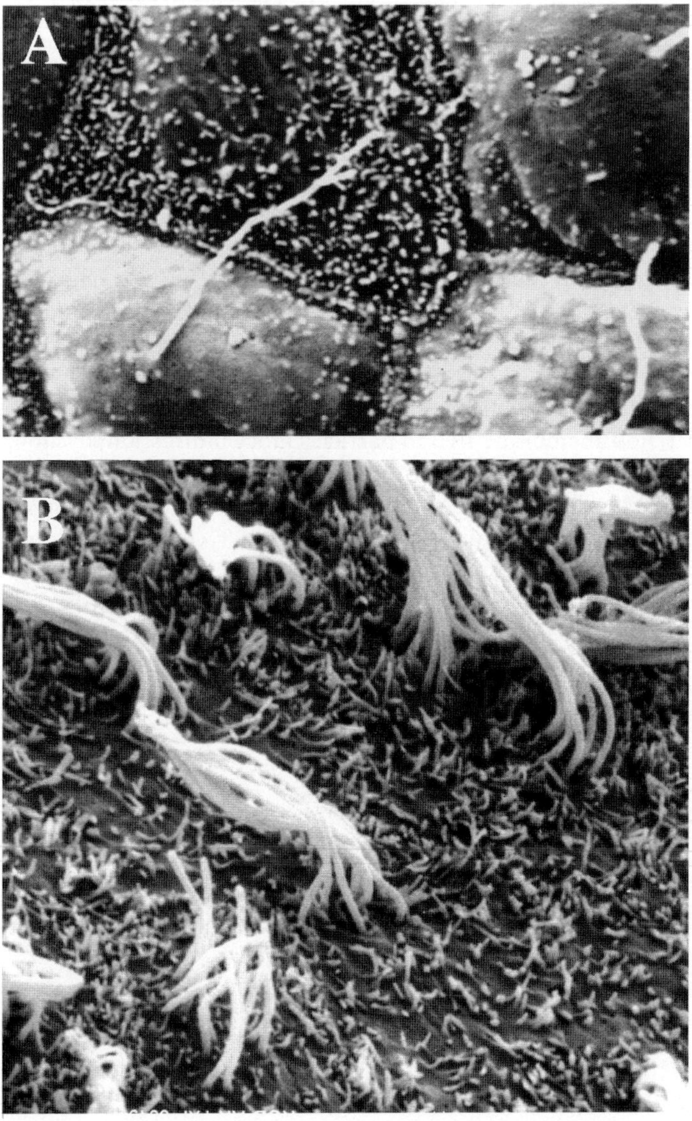

Fig. 17A, B Scanning electron micrographs of ciliated cells. **A** Monocilia in epithelial cells of kidney tubules (*arrowheads*). **B** Bunch of cilia in ependymal cells

and a variable number of associated intermediate, intermediate light, and light chains.

Kinesins are the molecular motors responsible for the anterograde microtubule-dependent transport in neuronal and non-neuronal cells. There are two types of kinesins, kinesin-I, which moves cargoes in neuronal axons and kinesin-II, which is involved in anterograde transport in motile and non-motile cilia. Kinesin-II has two forms heterotrimeric kinesin II and homodimeric kinesin II (Marszalek and Goldstein 2000). Heterotrimeric kinesin II consists of two distinct motor subunits, Kif3a and Kif3b/c and a third subunit called KAP, which is thought to bind the cargo during transport (De Marco et al. 2001).

Eukaryotic cilia and flagella are assembled through a process called intraflagellar transport (IFT). IFT was first discovered in the biflagellate alga *Clamydomonas reinhardtii* a free-living, unicellular organism that uses two long flagella for swimming and has proved crucial in the study of the biogenesis of cilia and flagella. It was in this organism that linear arrays of particles were first seen moving continuously from the base to the tip of the flagella (anterograde direction) and from the tip to the base (retrograde direction) (Kozminski et al. 1993).

This continuous movement of particles along the cilia is referred to as IFT. During IFT, non-membrane-bound particles move bidirectionally along the outer surface of flagellar microtubules from the basal body to the distal tip of the axoneme and back. These moving complexes include the molecular motor (kinesin or dynein) IFT particles, and the cargo, the latter consisting mainly of ciliary components that are transported from the site of synthesis to the growing tip of cilia. The IFT particle proteins, at least 17 polypeptides, were isolated from *Chlamydomonas* and are highly conserved in mammals including humans. So far two genes encoding IFT particle proteins, *Ift88* (also known as Polaris) and *Ift172*, have been identified as implicated in LR patterning in the mouse (Huangfu et al. 2003; Moyer et al. 1994; Murcia et al. 2000).

2 Glossary

Chemically induced mutation

A mutation induced by treatment with a chemical mutagen, for example, ethyl nitrosourea.

Chimeric embryo

Embryo formed by two different cell types. They can be created, for example, by combining two different early embryos (8-cell stage). Since tetraploid cells do not colonize the fetus, the aggregation between 2n (diploid) mutated stem cells and 4n (tetraploid) embryos ensures that the embryo is only formed by mutated cells, whereas the extraembryonic tissue is composed of otherwise normal tetraploid cells. This strategy avoids the extraembryonic requirement of a gene and allows analysis of its function in the embryo itself.

Chromosome region ("p" and "q" region)

The centromere divides each chromosome into two arms: the shorter arm, which is the p region, and the longer arm, the q region.

Enhancer

A DNA region that can increase the transcription of a gene. This region of DNA, usually found 5′ to the gene, functions as a control element that, when bound with a specific transcription factor, increases the level of expression of the gene. The difference between a promoter and an enhancer is that the enhancer is not sufficient to drive expression.

Fibroblast growth factors (Fgfs)

A fundamental group of secreted proteins with very important signaling functions during development. This is an ample family including more than 20 members that signal through 4-membrane tyrosine kinase receptors.

Gap junctions

Connections between cells that provide direct intercellular communication allowing the free passage of ions and metabolites up to 1kD in size. Gap junctions are constructed of connexins that assemble forming channels that connect apposed cells.

Gene locus

The specific position of a gene in a chromosome.

Gene targeting (knockout)

Generation of a mutant organism in which a specific gene has been completely eliminated. The mutation is designed and introduced into embryonic stem cells to substitute the normal allele by homologous recombination.

Homolog

A gene similar in structure and evolutionary origin to a gene in another species.

Hypomorph

A mutant allele that produces less than the normal amount or activity of the gene product.

Insertional mutation

When the chromosomic insertion of a foreign gene (transgene) causes the interruption of an endogenous gene. The transgene may help in identifying the disrupted gene but occasionally this is a very difficult task since the insertional event may have coupled genomic deletions or reorganizations. It may even be that the insertion disrupts a more or less remote regulatory element, rather than the gene itself.

Isoform

Very similar but different versions of a protein. Isoforms may arise by multiple mechanisms: different gene loci, multiple alleles, different subunit interaction, different splice forms, or different post-translational modification.

Morphogen

Secreted protein that diffuses and forms a concentration gradient across a field of responsive cells. Cells are able to read the concentration of the morphogen and behave accordingly to it.

Notch signaling

One of the principal signaling pathways that act during development. The Notch receptor is a single-pass transmembrane protein that upon activation by ligands (Delta and Serrate, also transmembrane proteins) is proteolytically cleaved to liberate its intracellular domain. The Notch intracellular domain combines with the cytoplasmic protein CBF1 and translocates to the nucleus, where it regulates the expression of target genes.

OMIM

Online Mendelian inheritance in man.

Organizer and its equivalents

The blastopore is the central structure of gastrulation in amphibians. The group of cells that form the dorsal lip of the blastopore are capable of directing cells to form axial structures and if transplanted, are able to induce a secondary axis. Because of its properties, the dorsal lip of the blastopore was called the organizer. Functionally equivalent structures are Hensen's node in avian embryos, the node in mouse and other mammal embryos, and the Kupffer vesicle in zebrafish. In chick and mouse the node forms at the most anterior part of the primitive streak.

Ortholog gene

The most closely related gene in a different species. Ortholog genes have extensive sequence similarity and often conserved function indicative of a common ancestor.

Transcription factor

A protein that binds to a cis-regulatory element (e.g., an enhancer) and thereby, directly or indirectly, influences the initiation of transcription.

Transforming growth Factors-β (TGF-β)

A superfamily of secreted factors with very crucial functions during development. They include BMPs, activins GDFs, and Vg1 families, all of them with some members involved in LR patterning.

Wnt signaling

One of the main signaling pathways during development. Wnt proteins are diffusible, extracellular ligands that interact with membrane receptors of the Frizzled and LRP families. The Wnt transducing signaling cascade controls gene expression by regulating levels of the intracellular protein β-catenin.

Subject Index